JESUS OF NAZARETH,
THE CHRIST

JESUS OF NAZARETH, THE CHRIST

A study guide to the Synoptic Gospels
for 'O' Level

by

P. R. TAYLOR, M.A., B.D.

*Head of Religious Education Department,
Kidbrooke School*

*Formerly Examiner in Religious
Knowledge, G.C.E. 'O' Level,
University of London*

HULTON EDUCATIONAL PUBLICATIONS

©
1964
P. R. TAYLOR

ISBN 0 7175 0455 7

First published 1964 by Hulton Educational Publications Ltd.,
Raans Road, Amersham, Bucks.
Reprinted 1966
Reprinted 1972
Printed Offset Litho in Great Britain
by Cox & Wyman Ltd., London, Fakenham and Reading

CONTENTS

	Page
Introduction	7

1. The Gospels
   A. *What were the evangelists trying to do?* — 9
   B. *The authority of the gospels* — 12

2. Son of Man and Son of God — 20

3. John the Baptist — 30

4. The Baptism and Temptations of Jesus
   A. *The Baptism* — 34
   B. *The Temptations* — 36

5. Galilean Ministry — 40

6. Jesus and the Other Religious Groups — 47

7. Miracles — 54

8. Parables — 62

9. The Kingdom of God — 74

10. The Sermon on the Mount
    A. *The problem of understanding the Sermon on the Mount* — 86
    B. *Certain presuppositions reconsidered* — 90
    C. *An analysis of the Sermon* — 94

11. The Disciples of Jesus — 100

12. The Confession and the Transfiguration
    A. *The Confession* — 108
    B. *The Transfiguration* — 110

CONTENTS

13. The Last Week
    Sunday — 115
    Monday — 117
    Tuesday — 120

14. An Interlude: The Little Apocalypse — 125

15. The Last Week
    Wednesday — 130
    Thursday — 130

16. Jesus on Trial — 138

17. The Resurrection of Jesus — 143

    Index — 148

INTRODUCTION

THIS book has been written to help examination candidates to acquire an accurate knowledge of the Synoptic Gospels and their relevance to today. It covers the major areas of the Synoptic Gospels in such a way as to prevent those frustrating questions, 'What can I say about this incident?' or, 'What has this to do with me?'

A careful reading, and re-reading of the Gospels themselves is a first essential. This is more difficult to achieve than is popularly assumed because pupils' familiarity with the text can be a handicap to a thorough study, and far too many candidates merely assume they know the text. The intention of this book is to present the ministry of Christ in a fresh way so that readers will be stimulated to think again about much that familiarity has deadened into a formal acceptance. For this reason statements are deliberately provocative and controversial in order to stimulate thought.

Apart from considering the needs of candidates who are following examination syllabuses and resting upon a firm basis of Biblical doctrine, the book takes into account recent advances in scholarship for those pupils intending to go on to more advanced examinations.

As the author is employed by the London County Council, it is necessary to state that the Council is in no way committed to the views expressed in this book.

Thanks are due to the following Examining Boards for permission to reproduce copyright material from papers set by them:

INTRODUCTION

The Associated Examining Board for the General Certificate of Education; The Senate of the University of London; The University of Cambridge Local Examinations Syndicate; Oxford and Cambridge Schools Examination Board; Oxford Delegacy of Local Examinations; Southern Universities Joint Board for School Examinations; University of Durham School Examinations Board.

P. R. T.

1

THE GOSPELS

A. What were the evangelists trying to do?

THE gospels are not biographies of Jesus. The writings of the evangelists cover only a short span of Jesus' life and personal details are lacking in their account. Therefore, to attempt to reconstruct the story or life of Jesus is rather futile. This is not to say that the study of the gospels is a waste of time. On the contrary, there is much value in attempting to appreciate how an author tries to explain his reaction to the historical fact that at a certain period of human history God intervened in the affairs of man by sending his Son, Jesus, to earth as a man. Here, then, is the fundamental problem confronting the evangelist: to attempt an understanding of the unique 'person' who is both man and the Son of God. Each evangelist brings with him to this problem certain dominant themes, certain peculiarities of thought and of method which must be examined carefully before we can appreciate Jesus as the Christ.

Each gospel begins at the end. To the evangelists Jesus is primarily the risen Lord, the person who overcame death by his resurrection. This final act in his life colours and gives meaning to many of the earlier events in his ministry. Jesus' power over death is foreshadowed in the raising of Jairus' daughter and the widow of Nain's son. If Jesus had power over life and death, then his control extends to the whole of nature: he is able to calm the storm. Only the Christ could feed the multitudes, thus anticipating the final Eucharist at the Last Supper.

It is equally true to say that the evangelists begin their study of Christ long before the birth of Jesus. For several hundred years the Jews had been expecting God to send His servant, or Messiah, to help them. Therefore Jesus must behave in the way stated in the Old Testament if they were to convince Jews that he was the Christ. However peculiar it sounds to modern ears, the evangelists thought they were giving evidence when they cited 'proof texts' from the Old Testament. 'This was done that the scriptures might be fulfilled', 'as it is written in the prophets', are not empty phrases for them. If this use of the Old Testament appears superficial, St. Matthew, in particular, can make much more subtle use of it. For him Jesus was the ideal representative of the true Israel, the Israel that obeyed God's will, and to him, therefore, Jesus in the course of his life had to undergo similar experiences to those which had occurred in Hebrew history. Matthew begins his book, 'The genesis of Jesus Christ'; Christ is taken into exile in Egypt, he passes through the water at his baptism, and the forty years in the wilderness are symbolized by his forty days of the Temptations. Moses received the Law on Mount Sinai, but Jesus reveals a new law during his Sermon on the Mount. Mark, on the other hand, is influenced much more by a particular passage in the Old Testament, Isaiah 52–53, so that one writer says that Mark's gospel is an extended commentary upon the 'Suffering Servant' passage in Isaiah.

Whatever method a particular evangelist used, he knew that he was writing his gospel to meet the real difficulties that Christians were facing in the Roman Empire. Therefore, the gospels have great practical value for the time at which they were written. One problem that worried Christians was that Jesus and his disciples were all

Jews, but the Church was now becoming predominantly Gentile. The evangelists explain, however, that there is nothing to fear in the inclusion of Gentiles in the Church, for Jesus himself had more than hinted at such a possibility when he spoke to the Syro-phoenician woman and healed the centurion's servant. Jesus had been aware of the destruction of Jerusalem in A.D. 70 (*cp.* Mk 13 and Matt 22[7]) as well as the rejection of the Jews (*cp.* Mk 12[1-12] and Matt 21[33-41]). Another difficulty that confronted the early Church, as it expanded, was the increase in the number of disciples, some of whom claimed to be true teachers of Christians. How could the true brethren be distinguished from the false, and the charlatans? When Jesus sent out the Twelve and the Seventy, he gave such explicit instructions that no doubt could exist in the minds of his hearers as to the conduct of a true Christian minister. Many similar problems are answered by the evangelist and we should be able to recognize them when they occur.

One does not, however, study the gospels out of either a literary interest in the Old Testament or a desire to learn more of Jewish history. The gospels are first and foremost a study of God's revelation to man of His will and purpose through Jesus Christ, His Son. As well as being a historical event this is also a present reality. Jesus is not dead, but is alive. He is still participating in the affairs of men in a variety of ways. 'Lo, I am with you alway, even unto the end of the world' (Matt 28[20]). Whether we have greater knowledge of God's will, or better means to implement it now, than men did nearly two thousand years ago, is a debatable point. What is indisputable is that the gospels contain eyewitness accounts of men who had been in personal contact with Jesus during his earthly ministry and, however the evangelists have dealt with this material, an enlightened study of their work can help us to appreciate Jesus as the Christ and to come to a better understanding of him today.

B. *The authority of the gospels*

Since the foundation of the early Church the gospels have held a special authority for Christians. For many centuries men claimed that this authority rested upon the fact that these writings contained the historical evidence of Christ's ministry, including the actual words he spoke, enshrining details of fact which must be true because all the accounts were directly inspired by the Holy Spirit. Each section of the gospels was given equal weight and importance. To criticize or doubt any particular statement was tantamount to blasphemy. There was considerable fear that once the literal truth of any part of the gospels was challenged the gospels as a whole would be of no account. In fact, a 'gospel truth' became something synonymous with an established and acknowledged statement which could never be criticized.

However, the modern critical spirit employed by men in every other walk of life to such good effect could not allow an unthinking acceptance of 'gospel truth' in religious experience. One line of inquiry sought to explain that historical truths are not self-evident. Life is made up of events which, when they are reported, are known as history. Events in history cannot be considered in isolation. Each event is only significant within its context. Men can perform identical acts for different reasons. Moreover, the experience and background of the historian reporting the events can colour and distort the 'historical truth'. A true historical event contains within itself not only the observable and reportable 'fact' but also emotional, volitional and spiritual values. To ignore these does less than justice to the historical situation.

Therefore, one must approach the gospels with caution. They do indeed contain a record of the historical facts of Jesus' ministry but the accounts have been coloured by the selection of these facts, their method of presentation, the particular interests of the author and the needs of the people who were going to read the completed accounts. All of these factors have to be weighed before extracting statements from the gospels and claiming that such statements are complete, absolute and final, valid for all men at all times. Of course the gospels are 'true', but their truth is relative and partial. The gospels are like precious gems. They reflect various facets of the truth but the whole or complete truth is beyond our grasp. Perhaps a more valid assessment of the gospels' importance is to say that they are *about* Jesus. A more comprehensive understanding of the truth is to know Jesus personally. The gospels may be one means to this end.

Evidence that the gospels are open to the same criticisms as other historical writings is apparent when they are studied carefully. The same incident can be used in a different context by two evangelists to illustrate entirely distinct and separate ideas (Mk 14 $^{3-9}$, Lk 7$^{36-50}$). A slight change of emphasis and the addition of a few details can alter the nuance of a parable (*cp*. Matt 22$^{1-14}$, Lk 14$^{15-24}$). Details differ in reporting an important event such as the Transfiguration. But surely this is what we would expect in any reporting. If everything tallied in three separate accounts written at different times, then we would suspect collusion and be left with the feeling that this was all too good to be true. Variations within the text of the gospels do not detract from but rather enhance their intrinsic truth.

A much more serious problem is the fact that at least twenty-five years elapsed between the death of Jesus and the appearance of a gospel in the present form in which we have it. If we date St. Mark's gospel A.D. 60–65, then what occurred during the preceding two decades to men's understanding of Jesus' ministry is of vital importance.

To assume, with certain scholars, that the Church during this period 'invented' a gospel, under the guidance of astute men like Paul, and that nothing can be known about Jesus or his ministry, is facile and takes no account of the historical facts.

In the first place much must have been written down before the gospels, as we have them, were produced. Sometimes it is difficult to remember that very many of the epistles were written before the gospels, because they are placed after the gospels in the New Testament. St. Paul, writing at the beginning of the second decade after the death of Jesus, bears witness to the preaching about Christ, as well as to an understanding of that material. In as much as this corroborates the evidence found in the gospels, we are more certain of our foundations. Again there is much support for those scholars who have traced an earlier written form for all three gospels in the Aramaic 'logia' of Matthew, and the Ur-Marcus and Proto-Luke hypotheses. Even if these ideas remain hypotheses and not established facts, there still remains the extremely strong possibility that the earliest Christian preachers gathered together certain texts from the Old Testament as witnesses to Jesus Christ and used these as a basis for their sermons. The evidence of such collections is so well established in the gospels themselves as to need little comment.

Many scholars have tried to probe beneath the surface and by a comparative study of the synoptic gospels to arrive at an understanding of a possible early written source. This 'document' they call Q from the German word *Quelle* meaning 'source'. As Mark wrote his gospel first, which was read by both Luke and Matthew, then both these evangelists could add their own material (L and M) to the Q source. Streeter was the first scholar to analyse the gospels by the following diagram:

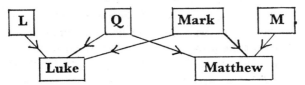

As helpful as such a theory is, it appears far too cut-and-dried to do justice to the complexities of the situation. The assumption that Q was one written source is far from proved. In fact, there could have been very many 'sources'. Furthermore, the material designated L and M could, and most certainly did, come from a variety of origins: friends and acquaintances of Luke or Matthew, who had known or heard Jesus during his earthly ministry, passed on information. The manner in which this information was used by the evangelist is another matter. A study of synoptic relationships will show that, even after the final selection of materials, each evangelist used the material in a different manner. One incorporates whole sections of Mark verbatim; the other rewrites and rephrases Mark and weaves it into his own account. Much useful and informative detective work can be done with a synoptic chart.

An attempt to reconstruct the 'original gospels' and their sources is as inconclusive as it is complex. In fact, the whole tenor of recent research is to demonstrate the intricacies and details confronting the scholar. What emerges from a vast amount of study is that no conclusions are drawn from guesswork or the opinions of individuals. All the work is a sincere attempt to arrive at some understanding of the truth. A nagging demand for certainty must never blind us to the needs of truth. Perhaps the present position with regard to gospel origins may be stated rather crudely as follows. The early Church formulated, very soon after Christ's ministry, a statement containing the essential facts concerning the life, death and resurrection of Jesus. Such a proclamation was often interpreted in the light of Old Testament testimonia. So meagre and thin was the outline proclamation of the ministry that the evangelists were left to gather together a variety of evidence from different sources, oral and written, and arrange this material around the outline. Therefore, the gospels represent something of a hotchpotch. Interest sometimes centres around a particular place, such as Capernaum, or a group of related parables or sayings; at other times, there is the more or less established framework of a chronological narrative, as in the Last Week. One modern writer has even gone so far as to assume that the search for the original gospels is pointless and that each incident reported within the gospels must be taken on its own merits so that each 'section', or pericope, of the evangelist's work represents an element of the truth of the gospel. That is, each pericope is taken as a gospel 'within the gospels'.

Another aspect of the authority vested in the gospels must depend upon the integrity of the authors themselves. Here there is some evidence which is apparent even to the casual reader. John Mark was very much in the centre of affairs in the early Church. His mother Mary had a house in Jerusalem which was used for Christian activities (Acts 12^{12}). John himself was associated with the missionary work of his uncle Barnabas and, in spite of the argument with Paul, he appears to have aided this apostle considerably in his later work (*cp*. Col 4^{10}, Phil v. 23). Mark's gospel, at certain points, betrays evidence of having obtained material from eyewitnesses (*cp*. 4^{38}, 6^{39}, 14^{51}). Tradition associates John Mark with Peter at Rome. Although many claim that this was merely to give the gospel apostolic authority, there is the evidence of the gospel itself, in its treatment of Peter's character, which adds considerably to Papias' statement that Mark wrote his book under the guidance of Peter. That Mark's gospel was intended for a Gentile audience is certain, in so far as he always explains Jewish terms or customs (15^{22}, 7^{11}, 7^3, etc.). Jesus is portrayed as the man of action and a fulfilment of the Suffering Servant of Isaiah 53.

Much less is known of Matthew personally. The gospel, however, throws considerable light upon the man. Matthew was a Jew writing for a Jewish audience. For him Jesus is the Messiah foretold by the Old Testament prophets. Whilst being keenly aware of the rejection of Jesus by the Jews, he is critical of allowing Gentiles into the Jewish-Christian Church too readily (*cp.* 7^4).

Luke, on the contrary, is very concerned with the universal appeal of the gospel, that it is intended for all men everywhere. Attention is particularly given to the social outcasts—sinners, tax-collectors, the Samaritans, and those underprivileged members of society at that time, women. The predominant impression left after reading this gospel is the reasonable, law-abiding and respectable nature of the Christian community. As Luke dedicated his work to the 'most excellent Theophilus', that is, the modern equivalent of 'His Excellency', a high ranking Roman official, several commentators have regarded the gospel as the first Christian 'apologia'. Yet Luke was certain that the gospel would not spread to all men by human agencies alone, so that behind all the events he sees the divine influence at work through the Holy Ghost. This is what one would expect of 'the beloved physician' who was so closely associated with Paul on his missionary tours that he kept a diary, which he later included in the Acts, and which is known as the 'we' passages.

What has been said above is not an attempt to 'prove' the authority of the gospels. Rather it is a warning that one needs neither to accept blindly the truth of the gospel nor to pick and choose according to an irrational whim, as to the veracity of a given section. The truth lies much deeper than this, and perhaps one can only grasp at a facet of any truth. In many subjects the answer is either right or wrong—for example, the solution of a simple problem in mathematics, or the conjugation of a verb in French. In studying the gospels, we are dealing with man's understanding of God in the total complexity of life. Life is not black or white, right or wrong. All is merged into dull shades of grey and it is only the bigot who sees his opponent as absolutely wrong. Perhaps one grave error men have committed historically is to assume that the gospels *alone* bear witness to the truth. The gospels are indeed a true witness to the fact of God at work amongst men but they are not the only witness to this fact. There is the fact of the experience of men, committed to Christ's cause throughout all the ages. This evidence of the worshipping community, commonly called 'the Church', cannot be isolated from the gospel witness. Moreover, man has had at all times the continual help of God through the activities of the Holy Spirit. To neglect any of these in order to enhance 'a gospel truth' is dishonest. Perhaps the only method of studying the gospels is to remember with all humility the words of St. Paul, 'For now we see in a mirror, darkly: but then face to face: now I know in part; but then shall I know even as also I have been known' (1 Cor 13^{12}).

POINTS FOR DISCUSSION

1. What evidence can you give from St. Mark's gospel of (*a*) his use of details obtained from eye-witnesses; (*b*) his use of Aramaic terms and their explanation; (*c*) his explanation of Jewish customs?
2. What examples can you find in St. Matthew's gospel of the following:
    (*a*) Fulfilment of Old Testament prophecy
    (*b*) An interest in the Last Days (or the Day of Judgement)
    (*c*) Hostility to the Roman occupation
    (*d*) An interest in the Jewish Law
    (*e*) Anti-Gentile propaganda?
3. 'St. Luke was concerned with the sinner and the social outcast.' Give examples of his interest in such people and discuss his reasons for writing a gospel.
4. How do the following incidents provide an answer to certain problems that concerned the early Church?
    (*a*) Parable of the Sower. (Mark 2^{2-20}.)
    (*b*) Visit to Nazareth. (Mark 6^{1-6}.)
    (*c*) Payment of tribute. (Mark 12^{13-17}.)
    (*d*) Great Supper. (Luke 14^{15-24}.)
    (*e*) Last Supper. (Luke 22^{14-20}.)
5. State briefly what you understand of the synoptic problem, illustrating your answer from the Gospels. (A.E.B., Summer 1960.)

2

SON OF MAN AND SON OF GOD

WHILST reading the gospels, one is always conscious of the dilemma presented by the character of Jesus. On the one hand Mark plunges boldly into his introduction, 'The beginning of the gospel of Jesus Christ, the Son of God'. Luke, on the other hand, is more cautious and, after describing the genealogy of Jesus, ends with, 'the Son of Adam, the Son of God'. The idea of a 'Son of Adam' is a simple one because it is so familiar. We are all sons of Adam. It implies a human being created with all its frailties and limitations. But the concept of 'the Son of God' is far from familiar. This implies an intimate connection with the creator, who has no limitations and is a being completely good, completely everything. The dilemma is apparent when one tries to combine the two ideas into one person.

That Luke was correct to describe Jesus as the Son of Adam is perfectly clear from reading the gospels. Jesus was born in the normal way to a woman and brought up in the atmosphere of an ordinary family. 'Is not this the carpenter's son? Is not his mother called Mary? And his brethren, James, and Joseph, and Simon, and Judas? And his sisters, are they not all with us?' (Mk 6[3], Matt 13[55]). Moreover, he must have developed roots in Nazareth, for his visit there is specially mentioned. To all outward appearances, he engaged in all the normal human activities. 'The Son of Man came eating and drinking, and they say, Behold a gluttonous man, and a winebibber, a friend of publicans and sinners.' Personal friends were an important part of his social activities and included a wide variety of people from Simon the Pharisee, Mary and Martha to Zacchaeus the tax-collector and Mary Magdalene. We see him experiencing the commonest human feeling of hunger during the temptations and thirst upon the cross and, on more than one occasion, so strong are his feelings that he weeps. There is nothing of the prig about this man. When one fellow addressed Jesus as 'Good Master', he was quick to reply, 'Why callest thou me good? None is good save one, even God.'

Yet there is a danger in describing Jesus, the perfect man, in these terms. The very terms and incidents used to demonstrate his character may become so idealized and so romantic as to divorce Jesus from the reality of the world we know. Ultimately we would end up with a figure lacking personality, a benign apparition walking round in a white sheet dispensing goodness to a world that he did not understand and which could not take him seriously. Several Hollywood religious epics have recently taken this point of view. Nothing could be further from the truth of the gospels. Jesus was above everything else the supreme realist. Many of his statements have a harsh, almost cynical touch. 'No man, having put his hand to the plough, and looking back, is fit for the Kingdom of God' (Lk 9^{62}). 'Many are called, but few are chosen.' 'To him that hath shall be given, but to him that hath not, even the little he hath shall be taken away and given to another.' Such a man acknowledged the necessity of facing up to reality and the acute struggle that this would involve. 'But no one can enter into the house of the strong man, and spoil his goods, except he first bind the strong man; and then he will spoil his house' (Mk 3^{27}). The thought of actual physical force was faced squarely when Jesus entered the Temple during the last week and forcibly evicted those buying and selling. Moreover, he warned his disciples of the fight to come and the necessity of being prepared to meet it: 'But now, he that hath a purse, let him take it and likewise his scrip; and he that hath no sword, let him sell his garment, and buy one' (Lk 22[36]). Perhaps one may argue, this is over emphasizing the rôle of actual crises and not looking at the average day-to-day Christian life. Yet even here Jesus realized the hardship and distress his gospel could cause, 'Think not that I came to send peace on the earth: I came not to send peace, but a sword. For I came to set a man at variance against his father, and the daughter against her mother, and the daughter-in-law against her mother-in-law: and a man's foes shall be they of his own household' (Matt 10[35-36]).

The above is not intended to present a new picture of Jesus' character as a hard bitten cynic but merely to gain some balance in an assessment of the term 'the Son of Adam'. There is nothing mutually incompatible in the tender and gentle Jesus playing with the children and the vehement and caustic rabbi putting the Pharisees in their place. It reflects part of the total human situation. It gives us a point of contact with Jesus and the knowledge that he is dealing with the same situation in which we find ourselves.

Jesus frequently refers to himself as the Son of Man. The Hebrew word for man is *adam*. The two phrases 'Son of Man' and 'Son of Adam' could be interchangeable and contain the basic idea of the most human of men. The phrase is used in this sense by the Psalmist: 'What is man that thou art mindful of him? And the Son of Man that thou visitest him?' (8[4]). To restrict the meaning of the phrase in this manner would do violence to the significance of the term to those who heard Jesus use it. A rather illuminating mention of the phrase is made in Daniel 7 when, after the collapse of all the large worldly empires, the Jewish nation will triumph. The Son of Man here is a collective figure representing the whole Jewish nation, whose ultimate end is glorification by God.

'One like the Son of Man came with the clouds of heaven, and came to the Ancient of Days and they brought him near before him. And there was given him dominion, and glory and a kingdom, that all people, nations and languages should serve him: his dominion is an everlasting dominion which shall not pass away, and his kingdom that which shall not be destroyed' ($7^{13, 14}$).

Without doubt many of the fundamental ideas of this passage colour the disciples' attitude towards Jesus. These are the arrival of a person from the clouds of heaven, having a kingdom, throned in glory and establishing a permanent régime. Matthew's conscious desire to prove Jesus the ideal collective Jew by making him undergo symbolically all the events of Hebrew history has already been mentioned. Consequently it is not surprising that Luke reports the disciples after the resurrection asking Jesus, 'Lord, dost thou at this time restore the Kingdom to Israel?'

There are further overtones of meaning in the phrase 'Son of Man' which would be familiar to Jesus and his contemporaries. In the Book of Enoch the 'Son of Man' is described as a semi-divine figure who is coming to judge the world and set all to right. Politically, the Jews at the time of Christ were powerless; there was little hope of establishing an autonomous government. The might of Roman power, coupled with the apparent hopelessness of any man to overthrow it, led the Jews to speculate about divine intervention through the person of the Son of Man.

Jesus' use of the term Son of Man, with reference to himself, may have been quite deliberate. Partly because the phrase was concrete and definite enough to have a specific meaning and partly because it was vague and amorphous enough to absorb and include within itself any other particular significance that Jesus wished to give it.

Jesus Christ is not a name in the same sense as Bill Smith. The meaning is made clearer if it is translated Jesus, the Christ. Christ then becomes an official title and not a surname. Christ is a Greek word, a synonym for the Hebrew word Messiah, which means 'to anoint' or 'the anointed one'. Anointing, that is, pouring or smearing of oil, is a common feature of the Old Testament. Jacob anointed the stone used as a pillow when he had a dream at Bethel which later became an official sanctuary; David the shepherd was anointed by Samuel to mark his appointment as King; Joshua and Zerubbabel are the anointed ones according to Zechariah when they took over the leadership of Jerusalem after the exile. Even a foreigner, the Persian King Cyrus, could become an anointed one, or the psalmist who says, 'Thou anointest my head with oil: my cup runneth over.' In each case anointing is a symbolic act, denoting God's favour and the recipient is set apart to perform a special service for God. Jesus, the Christ, had both of these attributes and much more; for, just as the idea of the Son of Man has a variety of meanings, obtained over a long period of time, so has the term Messiah. The difficulty is not to give the word a meaning but to ascribe the right nuance to it on a particular occasion.

Many scholars believe that a true understanding of the term Messiah can only be ascertained by a study of a ceremony of the enthronement of a 'King' during the early years of the Hebrew monarchy. This was not the actual coronation of the King but was an event held each year, when, in a desire to renew the fertility of the soil, to ensure good harvests in particular and the well-being of the people generally, the King acted the part of God's representative. The ceremony must have been a grand affair. Part of it included the chastising and abasement of the King, obviously as a form of penance. The climax would be reached when the King, completely purged, assumed his throne and was given all the regalia of office. Harmony had once more been restored between God and his people through the office of an individual and the divinely ordered government could proceed for the ensuing year. Undoubtedly evidence exists for such a festival and even a casual reader can appreciate the significance of some of the enthronement psalms. The actual abasement before the glorification of the King can be clearly discerned in Psalms 22 and 69:

> 'But I am a worm, and no man;
> A reproach of men, and despised of the people.
> All they that see me laugh me to scorn:
> They shoot out the lip, they shake the head, saying
> He trusted on the Lord that he would deliver him.'

Then comes the climax of the enthronement in Psalm 110 where each verse is accompanied by a suitable action of ascending the throne and receiving the regalia. One of the final statements is the glorification of God's chosen, bringing to mind all the promises made to David and his descendants in Psalm 89.

Many of the basic ideas behind such ceremonies as these are interesting and instructive for the light they shed upon the rôle of 'the Christ'. The Hebrews had already arrived at the idea of an individual being God's representative, and an individual who had to be despised and rejected before he could be glorified. An essential part of his work was redemption, so that God and man could be brought together and the results of such harmony seen in the everyday life of the Jew by God ratifying all the promises made earlier in his covenant. Jerusalem was the centre especially connected with these saving acts, and a 'Son of David', a royal person, the central figure. Any connection between these ideas and the work of Jesus is so obvious as not to need further explanation. But one idea is worthy of further comment. An important element in the work of the 'enthroned one' was his control over nature for man's benefit. It is precisely this point that John the Baptist stresses when he preaches about the one coming from God, 'Every valley shall be filled, and every mountain and hill shall be brought low: and the crooked shall become straight, and the rough ways smooth.' To modern ears these statements sound like poetic exaggeration. But a Hebrew living in the first century would be surprised if the Messiah could not exercise such control over nature!

Another very significant train of thought in Hebrew speculation about the Messiah was that it was necessary that he should suffer. To twentieth-century ears, this appears as objectionable as it is unnecessary. But, however distasteful suffering may be at a certain time, the ultimate results it produces can be worth while. On several occasions Jesus told the disciples of the necessity for his suffering. 'Behoved it not the Christ to suffer these things, and to enter into his glory?' (Lk 24[26]). Over five hundred years before the birth of Christ a man known as Second Isaiah had arrived at the conclusion that, if a man were to be obedient to God's will, he would be spurned and rejected by his fellow men. His thoughts on this theme are found in a series of poems included in the text of Isaiah 40–55. The best known poem is Chapter 53 and is commonly called 'the Suffering Servant'. Even a cursory reading will show how closely Jesus' life follows the pattern of Isaiah's servant. Whether Jesus saw himself as fulfilling this prophecy or whether the evangelists use this chapter consciously as a framework for Jesus' ministry is

SON OF MAN AND SON OF GOD

a moot point. The fact remains that Isaiah 53, in particular, has had a profound influence on Christian thinking about Messiah.

So far as Jesus is concerned he rarely speaks of himself as Messiah. He preferred to express his ministry in terms of a personal relationship with God, a father–son relationship. This is a particularly happy choice of image to describe a close and intimate association of two men. The father creates, cares for, and disciplines, or judges, his children. There is a blood kinship and yet the father is the dominant controlling partner. In no sense is there freedom of choice of a father. Jesus expressed this unity of aim as follows:

> 'All things have been delivered unto me of my Father: And no one knoweth the Son, save the Father; neither doth any know the Father, save the Son, and he to whomsoever the Son willeth to reveal him' (Matt 11[27]).

On many occasions during the ministry we see Jesus deriving strength by contact with God in both prayer and fasting. Over and beyond what Jesus says and does, there is an awareness of the ultimate power of God. His reply to James' and John's request for a privileged position in the Kingdom of God is instructive in this respect:

> 'but to sit on my right hand or my left hand is not mine to give: but it is for them for whom it hath been prepared' (Mk 10[40]).

When Peter suggested that there was no need for Jesus to suffer, he was rebuked with, 'Get thee behind me, Satan: for thou mindest not the things of God, but the things of men' (Mk 8[33]). Sometimes it was extremely difficult for Jesus to submit himself completely to God's will, as when he sweated blood in the Garden of Gethsemane: 'Father, if thou be willing, remove this cup from me' (Lk 22[42]).

The Sonship of Jesus was something that was particularly and peculiarly his alone. It is true Jesus taught his disciples to pray to 'Our Father ...' but it would be an impertinence to suggest that men stand in the same relationship to God as Jesus. It is only through Jesus Christ that we have the possibility of such a close relationship with God. So that the phrase at the end of a prayer, 'asking this in the name of Jesus Christ, thy Son', is not an empty phrase but very meaningful. St. Paul reminds us that we cannot consider ourselves as the natural Sons of God but as 'Sons of adoption'. In a legal sense it is Jesus who makes the adoption order possible.

Having looked at the implications of the terms Son of Man, the Christ, and Son of God, we are now in a better position to draw certain conclusions. All these terms are complex. A development in meaning extending over a lengthy period of time makes for difficult and almost archaic overtones. However, in the historical person of Jesus certain central ideas emerge. On the one hand we have a man, the most human of men, searching for security, peace and love in a world where the dominant theme is insecurity, tension and hatred. None of these attributes is sought for its own sake but from a belief that God intended a harmony to exist in His creation. Something had gone wrong between God and men; the 'Fall' was real. Therefore, on the other hand, there is the divine, royal figure having an intimate and unique knowledge of God's will who could provide men with a means of achieving harmony with God and His creation. The glory of God was revealed in obedience, service and suffering. But this was the Gospel, the joyous announcement, of Jesus Christ that there was now the possibility of reconciliation with God. The angels at his birth echo this sentiment:

'Glory to God in the highest,
And on earth peace among men in whom he is well pleased'
(Lk 2¹⁴).

POINTS FOR DISCUSSION

1. Basing your answer on St. Mark's Gospel, how would you answer a friend who asked you, 'Why was Jesus of Nazareth called Jesus Christ?' (Camb., Summer 1959.)
2. What evidence is there in the gospels, in the words and deeds of Jesus to suggest that he was no ordinary man?
3. What attitude did Jesus have towards Mary and his brothers and sisters? What teaching did Jesus give about the strength of family ties to his disciples?
4. Jesus is often called the Son of Man in the gospels. What do you think is the significance of this title?
5. Write brief comments upon the most important ideas contained in the following passages:

    (*a*) The Son of Man came not to be ministered unto, but to minister, and to give his life a ransom for many.

    (*b*) Behoved it not the Christ to suffer these things, and to enter into his glory?

    (*c*) For the Son of man is lord of the sabbath.

    (*d*) And Joseph also went up from Galilee, out of the city of Nazareth, into Judaea, to the city of David, which is called Bethlehem, because he was of the house and family of David; to enrol himself.

    (*e*) 'Truly this man was the Son of God.'

3

JOHN THE BAPTIST

JOHN the Baptist was a source of embarrassment to the early Church because he was so popular. Some were ready to accept him as the Messiah himself; others were prepared to recognize Jesus as John resurrected from the dead. The synoptic gospels do not seek to conceal his popularity, but tell us that 'all they of Jerusalem and Judaea went out to be baptized of John in Jordan', and even Acts mentions the disciple Apollos who went preaching his baptism alone. Consequently all the evangelists take considerable care in handling John's career. They make him describe his own efforts in derogatory terms and Jesus himself comments on the necessary shortcomings of John's ministry.

John's appearance was bound to create a stir since he was in the tradition of the Hebrew prophets. No one like him had been seen since the Exile. His wild appearance with his camel hair garb, his ascetic habit of eating only the bare necessities for life, his outspoken demands for moral purity, all these set against the background of the Judaean wilderness attracted attention. This wilderness was not a mere backcloth for his ministry but an essential part of it. Perhaps John would be understood better if the verse were punctuated, 'The voice of one crying, "In the wilderness, prepare ye the way of the Lord." ' The Jewish way of life had become too soft and sophisticated; Jews needed reminding of their past, particularly of their adherence to God's will during their privations in the wilderness. Other figures had emerged from it in times of crisis, figures like

Rechab in the time of Jehu. Unique in many respects, John's message had something in common with that of the monastic group in the South near Qumran called the Essenes.

The most startling news that John preached was the coming of Messiah. 'There cometh one who is greater than I the latchet of whose shoes I am not worthy to stoop down and unloose.' 'I indeed have baptized you with water but he will baptize you with the Holy Ghost.' Men had to prepare themselves for this great event by a complete change or renewal of their lives. The symbol John used to exhibit this change was baptism. Literally, baptize means 'to wash'. Such a custom was not unknown to the Jews; even the Pharisees practised it upon their Gentile converts. This was not Christian baptism, although Jesus did associate himself with John's work at his own baptism.

A note of urgency persists through all John's preaching, for this was a time of crisis. The choice had to be made now between good and evil. 'The axe is laid to the root of the tree . . .' If John was really going to prepare the way for the Messiah, then men must first recognize their moral obligations to one another. Luke's account of John's preaching expresses this demand for moral responsibility. The rich are to recognize the needs of the poor; tax-collectors and soldiers are not to take advantage of their official position to bully others. Once men 'brought forth fruits worthy of repentance', then the Messiah would have a firm foundation upon which to build his ministry.

John ardently believed that moral leadership should come from the top, that the royal family should set an example to society. Yet Herod, by associating with his brother's wife, was committing adultery. John's public attack upon Herodias was no private vendetta but part of a deliberate policy of purifying society. Ultimately, John lost his life for this cause, and in this clash of prophet and queen, one is meant to see a parallel with the Old Testament saga of Elijah and Jezebel. It is interesting to note that Herod, in spite of his wife's chicanery, showed a healthy respect for John and valued his judgements. 'For Herod feared John, knowing that he was a just man and an holy, and observed him; and when he heard him, he did many things, and heard him gladly' (Mk 6[20]). Although John was Jesus' cousin, and had a successful ministry, he hesitated in accepting Jesus as the Messiah. Whilst John was in prison, he sent some of his disciples to Christ, asking, 'Art thou he that cometh, or look we for another?' After giving them an answer by performing those deeds which only Messiah could do, Jesus dismissed these disciples and began to talk to the crowd about John. He praised him highly saying that he was greater than a prophet; and then, his limitations are revealed: 'Among those that are born of women there is not a greater prophet than John the Baptist: but he that is least in the Kingdom of God is greater than he.' John had shown a radical weakness in doubting that Jesus was the Christ.

JOHN THE BAPTIST

POINTS FOR DISCUSSION

1. In which ways did John the Baptist prepare the way for Jesus?
2. What did John the Baptist teach the people? State and comment briefly upon Jesus' opinion of John.
3. Comment briefly upon the most important points contained in the following statements:

    (*a*) But I say unto you, that Elijah is come, and they have also done unto him whatsoever they listed . . .

    (*b*) Whose fan is in his hand, and he will thoroughly cleanse his threshing-floor.

    (*c*) 'Suffer it now; for thus it becometh us to fulfil all righteousness.'

    (*d*) for Herod feared John, knowing that he was a righteous man and a holy, and kept him safe.
4. What is the relation between the ministry of Jesus and that of John the Baptist? (Durham, 1957.)

4

THE BAPTISM AND TEMPTATIONS OF JESUS

A. The Baptism

THE actual account of Jesus' baptism in all three gospels is extremely short. This must not blind us to the importance of the event. There is a large measure of agreement in the accounts. Jesus goes to John who baptizes him, the heavens open, a voice is heard declaring Jesus as God's son and there is the descent of the Spirit in the form of a dove. Behind these simple statements lies a deep and important significance for an understanding of Jesus' ministry.

By the act of going to John to be baptized, Jesus is making public two facts about himself. He directly identifies himself with all the other people who had been baptized by John and by so doing strengthens the ties between man, the Son of Man, and the Christ, the Son of God. Moreover, Jesus is showing his approval of, and his desire to be associated with, the drive for moral purity inaugurated by John. Mark is perfectly clear on these points, but the other two evangelists wish to minimize the significance of John, possibly because the baptismal movement was something of an embarrassment in the early church. Matthew makes John remonstrate with Jesus on the principle that the superior has no need of the inferior. 'I have need to be baptized of thee, and comest thou to me?' (Matt 3^{14}). John is silenced by the plea that this is

God's will and must not be hindered: 'Suffer it now; for thus it becometh us to fulfil all righteousness.' Luke, on the other hand, removes John from active participation by stating that the voice and the opening of the heavens occurred *after* Jesus had been baptized, as a result of Jesus' contact with God through prayer (Lk 3^{21}).

Not only are Jesus and John involved in this scene, but also God. The voice proclaimed, 'Thou art my beloved Son: in thee I am well pleased.' Echoed in this pronouncement are two Old Testament passages, Psalm 2^7 and Isaiah 42^1. Psalm 2 is an enthronement psalm and this passage occurs when the kingly figure is on the point of assuming his power. Therefore we cannot escape the implication that this event marks the beginning of a new reign and gains further significance from Jesus' first reported statement at Capernaum: 'The time is fulfilled, and the Kingdom of God is at hand: repent ye, and believe the gospel' (Mk 1^{15}). Isaiah 42^1 occurs within the context of the Servant Passages, the best known being the Suffering Servant of Isaiah 53. Jesus' baptism is here associated with his suffering. Right at the commencement of the ministry the evangelists stress where all of this will lead. That such a view is not fanciful can be seen by Jesus' reply to the request by James and John, in associating baptism with suffering. 'Ye know not what ye ask. Are ye able to drink the cup that I drink? or to be baptized with the baptism that I am baptized with?' (Mk 10^{38}).

The divine intervention was not limited to the voice for the heavens opened and a dove descended. Today we commonly ascribe to the dove the significance of peace. Biblically the image has much greater significance. The association of a dove, or bird, with water is clearly seen in the Noah saga in Genesis 8 where the release of a dove is the means of establishing the fact that there is a possibility of a new covenant between man and God. Jesus' ministry renews this possibility. Further there is the image in Genesis I of the spirit of God 'hovering' (like a bird) upon the face of the waters and God speaking in order to create. Little wonder is it that both of the constructive thinkers of the early Church take up the theme of recreation through Jesus Christ. John in the Fourth Gospel does not mention the baptism but begins with a hymn to the Logos, or Word of God, the creative principle that was with God from the beginning, 'and the Word became flesh' (John 1[14]). Paul repeats this theme in a different key by referring to 'the first born of all creation, for in him (Jesus) were all things created' (Col 1[15, 16]). Therefore, in many respects this symbolic act can be said to set the tone for the whole ministry.

Upon three occasions during Jesus' ministry does a voice bear witness to Jesus as the Son of God: at the baptism, the transfiguration, and beneath the cross. Each occasion is a crisis, and contains within itself something vital about the gospel. On the first two occasions it is God bearing witness to his son; the last time it is the Gentile Centurion (Mk 15[39]). Surely this is a most significant development?

B. *The Temptations*

Jesus was both the Son of God and man. As the Son of God, it was possible to assume that he could, and always would, do the will of God; but that as man he would be subject to the human weakness of opposing God's will. This conflict was particularly acute at the beginning of his ministry, whilst he was planning how he could best perform God's will upon earth. The incident, known as the Temptations, is an attempt of the evangelists to understand this conflict within the person of Jesus Christ.

Behind the account of the Temptations is a wealth of Old Testament material. Mark, in his typically concise manner, arrives at the heart of the matter when he says, 'and he dwelt with the wild beasts' (1¹³). The other man who dwelt with the wild beasts was Adam, but he had upset the divine harmony established by God in disobeying the divine command. Mark wishes to establish that another 'Adam' had come upon the scene but this time he would be obedient, performing the will of God perfectly and restoring the harmony between God and man. No longer was there need for the cherubim to guard God's creation from man but 'the angels ministered unto him'.

The testing of Jesus has its parallel in the testing of the whole Jewish nation in the wilderness with Moses. Jesus endured forty days and the Jews endured forty years in the wilderness; both were tempted to disobey God. Moses was commanded by God to take the people to Mount Sinai: Jesus was driven by 'the spirit' into the wilderness. Even the first temptation is reminiscent of God providing manna. Jesus' obedience is in contrast to the disobedience of the Hebrews.

Each temptation was a real and poignant experience for Jesus. He had just heard, at his baptism, the voice from heaven telling him, 'This is my beloved Son.' Now, doubts as to his divine sonship crowd in, as two of the tests begin, '*IF* thou be the Son of God'. The insidious nature of the command to turn the stone into bread becomes more real when one remembers the precedent of God feeding Elijah at the brook Cherith. Satan is repulsed by a statement from Deuteronomy 8³, the last section of which Matthew quotes in full (Matt 4⁴).

Not only is Jesus tempted to doubt his divine sonship but also to misuse his messianic powers. For he seeks to accomplish his task the more easily. Once more scripture comes to the aid of Jesus in refuting Satan. In return for power Satan demands that he should compromise with the world but this is challenged by a slightly amended form of Deuteronomy 6:13, 'Thou shalt fear the Lord thy God, and serve him.' Perhaps the climax of the event is reached when Satan himself uses the scripture to tempt Jesus into jumping from a pinnacle of the Temple. But it is a perverted use of scripture because Psalm 91 says, 'For he shall give his angels charge over thee, to keep thee in all thy ways. They shall bear thee up in their hands, lest thou dash thy foot against a stone' (vv. 11, 12). By omitting 'in all thy ways' Satan has transferred a statement applying to the day-by-day activities of man to a spectacular testing of God by man. Jesus' realization of this fact is shown by his quotation from Deuteronomy 6:16.

During his lifetime Jesus was tempted many times, after the so-called Temptations. Satan 'departed from him for a season', only to return later. A continuous pressure is put upon him to disobey God's will. On one occasion it is the leader of the disciples, Peter, who remonstrates with Jesus that he need not suffer, but Jesus replies, 'Get thee behind me, Satan, for thou mindest not the things of God but the things of man.' On another occasion, it is the religious leaders of the Jews who demand a positive sign of his messiahship and they are told that the only sign this generation will receive is the sign of Jonah. They did not understand. The climax of all Jesus' temptations was in the Garden of Gethsemane. He prayed that the bitterness of his suffering might be removed, yet 'not my will but thy will be done'.

POINTS FOR DISCUSSION

1. Describe the baptism of Jesus and discuss the significance that it held for him. (London, 1962.)
2. Discuss the importance and meaning of our Lord's Temptations. (Ox. and Camb., 1960.)
3. Answer the questions on the following underlined phrases in the passage:

> 'And straightway coming up out of the water, he saw the heavens rent asunder, and the Spirit as a dove descending upon him; and a voice came out of the heavens, "Thou art my beloved Son, in thee I am well pleased".'

(a) 'heavens rent asunder.' On what other occasion in the gospels is there mention of something being rent? What importance do you attach to this?

(b) 'the Spirit as a dove.' Why a dove? What does this remind you of in the Old Testament? On what other occasion did Jesus say, 'The Spirit of the Lord is upon me'?

(c) 'a voice.' On what other occasions in the gospels are voices heard bearing witness to Jesus?

5

GALILEAN MINISTRY

WHEN Jesus began his public ministry in Galilee by proclaiming 'the good tidings of the Kingdom of God', it caused great consternation among the people. There was a marked contrast between him and the ordinary rabbis, 'And they were astonished at his teaching: for he taught them as having authority and not as the scribes' (Mk 1^{22}). It was all so different from what they had known before that they began to speculate, 'What is this? A new teaching!' (Mk 1^{27}). So great was the interest aroused that 'there went forth a rumour concerning him into every place of the region round about' (Lk 4^{37}). An embarrassing situation was created for Jesus because it was not only in Capernaum that 'the house' was crowded out but also 'a great multitude from Galilee followed: and from Judaea, and from Jerusalem, from Idumaea, and beyond Jordan, and about Tyre and Sidon' (Mk 3^8). Huge crowds restricted Jesus' mobility and hindered him in his avowed task of preaching the gospel to as many as possible. Therefore he rises early in the morning 'a great while before day' (Mk 1^{35}) and moves on. He must have been thankful for the rest and seclusion offered by a journey across the lake by boat but even this had its disadvantages (Mk 6^{33}).

The reason for the people's astonishment is not hard to discover. Luke reports (4$^{16-30}$) upon a sermon in the synagogue at Nazareth which is so forthright and forceful that it creates an immediate challenge to the congregation. Jesus was handed the scriptures and read from Isaiah 61$^{1-2}$.

GALILEAN MINISTRY

'The Spirit of the Lord is upon me,
Because he anointed me to preach good tidings to the poor;
He hath sent me to proclaim release to the captives,
And recovering of sight to the blind,
To set at liberty them that are bruised,
To proclaim the acceptable year of the Lord.'

'Today, hath this scripture been fulfilled in your ears.' To many, who had known Jesus from his birth, this statement came as something of a shock. 'Is not this the carpenter?' (Mk 6³). But there was no denying that one had to recognize a power that was unusual. At Capernaum, after the healing of the paralytic, 'they were all amazed, and glorified God, saying, "We never saw it on this fashion!"'

The man was a complete enigma, particularly in his behaviour. Whilst making claims of a messianic nature he associated with people known to have bad reputations. The publicans were the lowest of the low, men who could make a living by collecting taxes, at extortionate rates, from their fellow countrymen for the benefit of an occupying power. Tax-collectors were his friends! 'Why eateth your Master with publicans and sinners?' (Matt 9¹¹). Jesus replied, 'They that are whole have no need of a physician, but they that are sick. But go ye and learn what this meaneth. I desire mercy, and not sacrifice: for I came not to call the righteous but sinners.' Here there is no sense of wishing to condone the wickedness or evil that may be quite apparent but rather to influence those generally considered social outcasts to find a new dignity through the friendship of Jesus. An illustration of this happening is found in Zacchaeus. Moreover, Jesus did not *only* associate with the disreputable. Simon, the Pharisee was eminently respectable, as can be judged by his remark about Jesus, when a prostitute was washing Jesus' feet, 'This man, if he were a prophet, would have perceived who and what manner of woman this is which toucheth him, that she is a sinner' (Lk 7³⁹). Normal conventional barriers are broken down when there is a sincere and honest approach to Jesus. This was recognized by Jesus, 'Her sins, which are many, are forgiven; for she loved much: but to whom little is forgiven, the same loveth little' (7⁴⁷). Christ did not come to establish a mutual admiration 'society' for the righteous but to allow *all* men the possibility of salvation. The inclusive nature of God's Kingdom certainly shocked the 'respectable' when they realized the fact that it contained people like 'Mary, that was called Magdalene, from whom seven devils had gone out' (Lk 8²).

No wonder that even his friends thought that he was insane! 'For they said, "He is beside himself"' (Mk 3²¹). Others ('scribes', Mk 3²²ff., 'Pharisees', Matt 12²⁹ff., 'some', Lk 11¹⁵ff.) thought that this power was diabolical and that Jesus was in league with the devil. 'He hath Beelzebub, and by the prince of devils casteth he out devils.' Jesus refuted this suggestion by stating the principle that his work involved a weakening of evil power over men's lives, and that if the evil powers were divided against themselves they could not last long. Such an exercise of power on his part involved a real struggle. However, men have misunderstood this struggle if they view it merely as the power of an exorcist. Many had the power to subdue evil spirits. Jesus was not preparing for the Kingdom of God by getting rid of evil spirits. The Kingdom of God came with Jesus. The Kingdom of God and the evil powers are mutually incompatible. 'But if I by the finger of God cast out devils, then is the Kingdom of God come upon you' (Lk 11²⁰). Moreover, Jesus not only casts out the evil spirits but also wants men to accept the positive values of the Kingdom of God. Once the demon has been expelled, God must take vacant possession. 'In modern language the old obsessions, complexes, morbid fears and desires, and so on must be replaced by new loyalties and affections. The man who has been delivered must give himself body and soul, and spirit, to God and the service of His Kingdom' (T. W. Manson, *The Sayings of Jesus*). Otherwise 'the last state of that man becometh worse that the first' (Lk 11^{26}).

There was the possibility of a misunderstanding and a large measure of disagreement between Jesus and the authorities over his treatment of the Sabbath. That a crisis arose over this issue is clear from the evidence cited of men being planted in synagogues and private homes on the Sabbath to see if Jesus would cure them. A close surveillance was kept upon Jesus and leading questions asked when an infringement of the regulations occurred. To appreciate the true significance of the situation one must remember that Jesus never criticizes the practice of Sabbath observance itself. As a person brought up according to Jewish traditions he must have appreciated the sanctity and value of such a rite. What he does criticize is the inability to see that a greater demand than any specific religious observance might be made upon men at that particular time. The Kingdom of God had come to men, time was pressing, evil powers were at work all the time. Therefore, the battle against these powers must proceed all the time. Periods of relaxation to observe either the Sabbath or fasting were out of the question. The demands of the Kingdom were greater. For this reason he could say, 'The Son of Man is Lord of the Sabbath' (Lk 6^5). By doing nothing, a positive evil may result: 'Is it lawful on the Sabbath day to do good, or to do harm? To save a life, or to kill?' (Mk 3^4). Every humanitarian instinct is offended by a rigid adherence to a rule whatever the consequences; 'What man shall there be of you, that shall have one sheep, and if this fall into a pit on the Sabbath day, will he not lay hold on it, and lift it out? How much then is a man of more value than a sheep?' (Matt 12^{11}). But what particularly angered Jesus was the knowledge that the Jews dispensed with strict observance of the Sabbath when any Temple business was involved. He was invoking the same principle in favour of God's business: 'Or have you not read in the law, how that on the Sabbath day the priests in the Temple profane the Sabbath, and are guiltless? But I say unto you, that one greater than the Temple is here' (Matt 12^5).

We have arrived at the heart of the matter. By abrogating certain religious observances, Jesus was proclaiming the presence of the Kingdom and asserting special privileges for himself by virtue of his close relationship with the Father. When challenged on the point of fasting, he admits as much, 'Can the sons of the bridechamber fast while the bridegroom is with them? As long as they have the bridegroom with them, they cannot fast. But the days will come when the bridegroom shall be taken away from them, and then will they fast in that day' (Mk $2^{19, 20}$). This the authorities could not accept. Indeed by stressing the fact that they believed Jesus was acting on evil impulses, they committed the ultimate blasphemy of denying God; 'Whosoever shall blaspheme against the Holy Spirit hath never forgiveness, but is guilty of an eternal sin. Because they said, "He hath an unclean spirit"' (Mk $3^{29, 30}$).

From what has been said about the Galilean ministry, one may gain the impression of rather a disastrous beginning to Jesus' work. Persistent hostility combined with thoughtless obstruction to hinder Jesus at every turn. But this is not a complete picture. The hostility, the hindrance and even the rejection of Jesus do not mean the defeat and end of God's purpose—the death of the Son did not mean that. Final success is assured in spite of temporary setbacks. Mark 4 establishes this point most clearly. Much seed is wasted whilst the sower is about his task, but ultimately there will be a harvest, some seed yielding a particularly high return. The lamp, whatever use it is put to, will in the end give its light when placed in its stand. Treasure cannot be concealed for ever; one day it will be revealed. The farmer knows the consternation of waiting for the harvest, all the setbacks of weather and nature, but at last he will harvest the crop. If the principle of life is contained in the seed, even in such a minute seed as the mustard, it will at length produce the 'great branches'. Nothing can thwart the ultimate success of God's purpose. The chapter ends with the disciples themselves fearing for their lives because of the storm. Human hostility is replaced by the possibility of a natural disaster. Jesus rebukes them and says 'Peace, be still.' 'Who then is this, that even the wind and the sea obey him?' Obviously the evangelist knew the answer.

POINTS FOR DISCUSSION

1. 'Jesus' Galilean ministry was marked by a growing hostility upon the part of the officials.' What were the causes of this hostility? How did Jesus overcome it?
2. Describe carefully Jesus' visit to Nazareth. Why was Jesus not surprised at their rejection of him?
3. Explain fully Jesus' clash with the authorities over (*a*) working on the sabbath and (*b*) casting out demons by Beelzebub.
4. Tell the story of the healing of the man called Legion in the country of the Gadarenes. Which points in this story are different from any similar healing performed by Jesus?
5. Jesus 'taught as one having authority and not as the scribes'. Explain and illustrate this. (Camb., 1958.)
6. Write an account of the feeding of the four thousand and of the conversation with the disciples about it afterwards. What differences are there in the story of the previous feeding? What significance have these miracles for the Christian Church? (Ox. and Camb., 1959.)

6

JESUS AND THE OTHER RELIGIOUS GROUPS

DURING his ministry Jesus either came into direct contact with, or was influenced by, other religious groups which existed in Israel. Scribes, Pharisees and Sadducees are the best known of these groups. In so far as the Pharisees, in particular, are frequently mentioned in the gospels one could easily make the false assumption that these groupings correspond roughly to modern political parties—that is, that all Jews must have been either sadducean or pharisaic—Conservative or Labour. Such would be a totally false impression. These groupings were not parties in the modern sense of the word. The total number of people included in these groups, at a most liberal estimate, was only seven per cent of the total population: ninety-three per cent were ordinary Jews. Yet the influence of this seven per cent cannot be measured in terms of their numerical significance so much as in the force of their ideas and the political, religious and social influence they were able to exert.

The origins of the Pharisees are somewhat obscure. They emerge in Israel as a coherent body, after the period of the Maccabean Wars, *circa* 150 B.C. As a group, they had distinct affinities with the Chasidic party which supported Judas in his struggle against Antiochus IV. The Chasidim were men enthused with a religious fanaticism for the purity of their belief. To achieve such purity they cut themselves off from any part of the community which did not share their views. Therefore many find in the name 'Pharisee' a clue to the origin of the group, that is, the 'Separatists' or those that separated themselves. Undoubtedly many Pharisees would have welcomed this description of themselves from what we know of their exclusive nature. Whether this is the true origin of the name is open to doubt. Many assume that their distinctive beliefs about the good and evil forces that control the world were derived from a study of the Persian philosopher Zoroaster. Therefore 'Pharisee' could be a nickname for the 'Persian sect'. Even today we refer to Christian groups by the place of the origin of their beliefs. We talk about members of the *Roman* Catholic or *Greek* Orthodox churches.

Josephus, the Jewish historian, informs us that there were about six thousand Pharisees in the time of Herod the Great. To this must be added their families and those who were not yet full members of the pharisaic group. So that one arrives at a figure of about twenty-five thousand people who, at one time, came under direct pharisaical influence, out of a total population of eight hundred thousand. Pharisees were laymen, being drawn mainly from the artisan and skilled worker classes. Small as they were in number, they were significant in influence. Credit for this must go to their organization. They met as a chaburah (club or brotherhood) with well-defined rules and officers. Usually they were closely associated with an area and provided great support for the synagogues. Leadership of the movement was provided by Rabbinical scholars, or the scribes. Sometimes they are called 'the scribes of the Pharisees' in the gospels. Therefore, this group was extremely well organized, close-knit and capable of decisive and effective action by dint of good leadership.

There could be no better witness to the Pharisees' fervent struggle to achieve righteousness than that of Jesus himself. They tried to live up to the command in

Exodus 19⁶ and they saw the divine providence working this out in history. They were conscious of the necessity of observing the will of God. In so far as they were able to fulfil God's will here on earth, the greater they thought their reward would be in a life after death. But they were willing to recognize the power of evil at work in the world and formulated much teaching on demons and angels. To combat the evil and to be secure in the knowledge of obedience to the divine will, the Pharisees were concerned to adhere strictly to the scriptures and to the 'Traditions', that is, the haggadah (explanations of matters of belief) and the halacha (explanation of matters of conduct). Much of this material was handed down in oral form.

Pharisaism represented progressive and radical thought in Israel. Josephus says Pharisees were always ready to indulge in political intrigue, invariably hostile to the government and even capable of sabotage and armed rebellion. Inasmuch as they saw the divine hand at work in history, their 'religious' activities extended to such fields as politics, economics and diplomacy (*cp*. Mk 12¹³⁻¹⁷). They were in the centre of affairs and Jesus rightly refers to their influence as the 'leaven of the Pharisees'. Amongst their number they included such liberal and scholarly Rabbis as Hillel whose statements in the 'Sayings of the Fathers' are as profound as they are enlightened. Christianity made a great appeal to many in the pharisaic ranks (*cp*. Acts 15, and Paul's position). Therefore, it is not surprising to see that Luke reports Jesus dining with Pharisees on no less than three separate occasions (7³⁶, 11³⁷ and 14¹). This friendly, if not very sympathetic, attitude of Jesus towards part of the sect can be seen in his reply to a scribe in Mark 12³⁴, 'Thou art not far from the Kingdom of God.' Consequently one must guard against making the assumption that all the Pharisees were 'bad' and all the Christians 'good'.

Nevertheless, it is on record that Jesus made some very

harsh statements about the pharisaical way of life (*cp.* the whole section in Matt 23). These objections rest upon a fundamental difference of opinion. To the Pharisee the Torah, or law, was all important. Now the Torah was not just a set of rules or regulations found in the scriptures, or comments found in the Traditions. Torah was the will of God given to man in diverse ways; as such it was sacrosanct. The supreme duty of all men was to obey the Torah. Jesus himself commended this struggle to achieve righteousness, 'The scribes and the Pharisees sit on Moses' seat: all things therefore whatsoever they bid you, these do and observe' (Matt 23[2, 3]). What Jesus objected to was their self-righteousness, their smugness in assuming that they always could, and did, obey the Torah perfectly. This is why the same passage continues, 'but do not ye after their works: for they say, and do not. Yea, they bind heavy burdens and grievous to be borne, and lay them on men's shoulders: but they themselves will not move them with their finger.' It must be clearly understood that Jesus is not saying that the Pharisees make false statements, or that he can give men a better theology than they can. Jesus' position is much more positive, and realistic. Stated crudely, it runs as follows: Rules of conduct are good, possibly essential for man. The struggle to do God's will is laudable. But man cannot achieve perfection on his own. He needs help from God, if only because human statements are limited and cannot cover the whole range of human activity. What is required is a closer relationship with God, conditioned by love. Such a relationship can only be realized through Jesus himself so that he can say to men, 'Come unto me, all ye that labour and are heavy laden, and I will give you rest. Take my yoke upon you and learn of me: *for I am meek and lowly in heart:* and ye shall find rest unto your souls' (Matt 11[28, 29]). Unfortunately the pharisaic attitude is still with us. There are

JESUS AND THE OTHER RELIGIOUS GROUPS 51

many professing Christianity who think they know all the answers, so that Christ has very little part to play in their lives. They are not sinners or bad men; they are good men who are earnest to do God's will.

'Except your righteousness exceed the righteousness of the scribes and Pharisees, ye shall in no wise enter into the Kingdom of heaven'(Matt 5^{20}). The operative word here is 'exceed'. True righteousness involves the whole man in a commitment to God. Excessive attention to any particular aspect of right-doing or an undue awareness of the impression one is making upon one's fellow men is sheer foolishness. 'But all their works they do for to be seen of men: for they make broad their phylacteries, and enlarge the borders of their garments, and love the chief seats in the synagogues and the salutations in the market places, and to be called of men, Rabbi' (lit: 'my great one') (Matt 23^{5-7}). Habitual smugness leads to self-deception. 'For a pretence [they] make long prayers' (Mk 12^{40}) and condemn themselves, 'God, I thank thee, that I am not as the rest of man, extortioners, unjust, adulterers or even as this publican' (Lk 18^{11}). Where the commandment governing outward acts was known, the Pharisees were impeccable; they even went so far as to tithe vegetables! (Matt 23^{23a}). But their strict observance of the letter of the law blinded them to the spirit that had inspired such a law. At times they were guilty of a lack of humanity (*vide* Corban incident, Mk $7^{9ff.}$). They 'left undone the weightier matters of the law, judgement, and mercy, and faith' (Matt 23^{23b}). These shortcomings prevented them from judging a particular case on its own merits. Perhaps their most damning feature was that although they had the ability and knowledge to lead men to a better knowledge of God's will they failed abysmally to do so. 'Ye took away the key of knowledge: ye entered not in yourselves, and them that were entering in ye hindered' (Lk 11^{52}).

The Sadducees, so far as the gospels are concerned, only came into contact with Jesus during the Last Week. No record exists as to their precise number, but they must have been a small and exclusive group. Many suggestions have been made about the origin of the word 'Sadducee'. Some relate it to Zadok, the priest, and see the Sadducean group as a priestly clique. This, however, does not tally with the facts. Not all Sadducees were priests, and certainly many of the priests and Levites were not Sadducees. Others regard the term 'Sadducee' as a flattering title meaning 'the righteous', from the Hebrew root S d k. A much more likely origin for the term is found in the Greek word which has come into the English language as syndicate. These men were part of that 'syndicate' or small ruling group who organized the legal, fiscal and diplomatic affairs of the Jewish state. Their headquarters was Jerusalem, their court the Sanhedrin, and their chairman the High Priest. Aristocrats by birth and rulers by nature they formed the élite of Hebrew society, wealthy men, who by modern standards, were 'realistic' in their outlook. In so far as their position depended upon the maintenance of law and order, they were the most reactionary and conservative element in society. Their frequent compromises with the occupying Roman authorities made them suspect so far as the common people were concerned.

The Sadducees' religious beliefs reflect their social position. Their tenets allowed greater freedom and initiative to the individual. Only the Written Law was acceptable. God created all the good, but the evil was the free choice which was presented to man. No intermediary elements such as Fate, demons or angels were considered necessary. The individual controlled his own destiny in life. They did not believe in a life after death, presumably because this life was the only one worth enjoying. In fact

JESUS AND THE OTHER RELIGIOUS GROUPS

it was a thoroughly worldly philosophy, commendable to a worldly ruling aristocracy.

POINTS FOR DISCUSSION

1. Write a summary of the origins of the Pharisees and Sadducees. What distinctive beliefs did each group have? Give examples of these from the text of the gospels.
2. For what reasons did the Pharisees criticize Jesus? Illustrate your points from the text of the gospels.
3. Why was Jesus critical of the Pharisees?
4. Distinguish between the Temple and a synagogue. Narrate an occasion when Jesus clashed with the Pharisees in a synagogue and the Sadducees in the Temple.

7

MIRACLES

Jesus taught by performing miracles. Such a statement is nonsense if miracles are regarded as supernatural events merely performed to cause a feeling of wonder and astonishment. It was a miracle in this sense that Jesus refused to accomplish when he was tempted to throw himself from a pinnacle of the Temple during his temptations. Of course miracles are not natural. In fact they could be described as events which contravene the known laws of nature. But all the gospel miracles occur within a natural framework by restoring or adding to those elements which are regarded as normal. For example, life is given to the dead, sight to the blind, health to the sick or peace to the storm. All of this presupposes a power and authority over and above that of which man is capable. This is just the point at which Jesus of Nazareth, the Christ becomes relevant. As both the Son of Man and the Son of God he had this power to perform miracles. Once this fact has been grasped it is pointless to inquire further into how miracles were possible. The more relevant question is why they were performed.

An answer to the question why Jesus performed miracles is given by Mark on the occasion of the healing of the paralytic man (Mk 2^{1-12}). 'Certain of the scribes' grumbled when they heard Jesus tell a man that his sins were forgiven. This was blasphemy in that Jesus of Nazareth was claiming to have the power of God. When Jesus became aware of their complaint he did not immediately cure the man of his paralysis and then advance this as proof of his power. Instead he questions the scribes and quite clearly states why he will heal the man: 'That you may know that the Son of Man hath power on earth to forgive sins.' Jesus had the right to intervene in the spiritual affairs of men. This so amazed the onlookers that they exclaimed, 'We never saw it on this fashion.' Many early commentators looking at this passage have stressed that the man was cured because of his faith. This is a rather awkward way of stating a truth. If Jesus performed the miracle to reveal something about himself, then there must be some response on the part of those viewing the miracle to understand its significance. Such a response is faith and this is why an element of faith in the beholder is always stressed. But the miracle was not performed because of the faith of either the men carrying him or the paralytic; Jesus could have performed the miracle without any assistance from them. Their faith made the miracle worth while. Mark leaves no doubt in our minds whose power actually performed the miracle. Jesus was not mobbed and fêted, but all were amazed '*and glorified God*'.

Another occasion upon which Jesus gives us an insight to his understanding of the nature of miracles is when John the Baptist's disciples visited him and asked, 'Art thou he that cometh, or look we for another?' According to Luke (7^{21}), Jesus performed many healing miracles and then said, 'Go your way, and tell John what things ye have seen and heard; the blind receive their sight, the lame walk, the lepers are cleansed, and the deaf hear, the dead are raised up, the poor have good tidings preached to them.' This reply echoes several Isaianic prophecies (including 35^5, 61^1), and this can be taken to mean that Jesus sees his miracles as fulfilling the Old Testament prophecies and inaugurating the messianic age. In this respect it is noteworthy that Jesus himself specifically connects the preaching of the good tidings (gospel) with the miracles. Miracles then become part of the means of revealing the Kingship of God. They are teaching. But this teaching is not meant to engender wonder but repentance and action. An instructive example of the influence of miracle is seen in the reaction of blind Bartimaeus (Mk 10[46ff.]): 'And straightway he received his sight, *and followed him in the way*', that is, as a disciple. The intimate relation between miracle and repentance is commented upon by Jesus in Matthew 11[20ff.]: 'Then began he to upbraid the cities wherein most of his mighty works were done, because they repented not. Woe unto thee, Chorazin! Woe unto thee, Bethsaida! for if the mighty works had been done in Tyre and Sidon which were done in you, *they would have repented long ago* in sackcloth and ashes.'

A call to repentance and action are only part of Jesus' teaching by miracle. As has been already said Jesus was, above all things, for the evangelists the risen Lord. One modern writer has even gone so far as to say that the Christian faith stands or falls by the last forty-eight hours of Jesus' life. Such a statement may be exaggerated but it does stress the vital importance of the last week. By healing Jairus' daughter Jesus is looking towards his own resurrection, and demonstrating that God has complete control over life and death. The whole episode is nicely constructed. On three separate occasions the certainty of the girl's death is stated. When Jairus first approaches Jesus he says, 'My little daughter is at the point of death.' The messenger from Jairus' house reports, 'Thy daughter is dead, why troublest thou the master any further?' At the house the mourners are scornful of Jesus' remark, 'The child is not dead but sleepeth.' The fact of the child's death is well established but, to Jesus, the finality of this state was no more than sleeping. The importance of the occasion is emphasized by the selection of only Peter, James and

John to witness the miracle. After the girl was resurrected, absolute secrecy was demanded because this was one of those events which would gain real significance after Jesus himself had risen. Just as after his own resurrection, as reported in Luke 24[41-42], Jesus asks for something to eat, he commands that the girl be given food. It would be the height of bathos to suggest that both the girl and Jesus felt hungry after such an experience! What the evangelists intend to demonstrate is the normal physical behaviour of both persons after the resurrection. There can no longer be the slightest suspicion that they are mere phantoms or apparitions—ghosts do not eat.

It is no accident that the account of Jairus' daughter is interrupted by the healing of the woman with the issue of blood. These events supplement the teaching of the other miracle by explaining the significance of the resurrection. This unfortunate woman suffering from haemorrhage was the symbol for the unclean. Uncleanness was the quality of not being whole. If a person was unwholesome he was not fit for the service of God. As uncleanness was contagious, it was forbidden to touch a dead body (Num 19[11ff.]), a leper (Lev 13), or a woman with an issue of blood (Lev 15[19ff.]). Although the woman had attempted to rid herself of her uncleanness by a variety of methods, all had proved useless; 'she had suffered many things of many physicians, and had spent all that she had, and was nothing bettered, but rather grew worse' (Mk 5[26]). In a last attempt she came and touched Jesus. Where all else had failed, Jesus succeeded. She was no longer unclean, but had been made whole. In these two incidents Jesus had broken down the traditional Hebrew barriers to God in those unfortunate cases of the unclean—the woman with the issue of blood and the dead girl. 'Unclean' now has no significance. So many incidents relate to the healing of lepers in the gospel that further comment is unnecessary.

All men, whatever their condition, can come to God through Jesus Christ. This is emphasized in Mark's teaching, when at the death of Christ, the veil in the Temple was rent, making access for all men easy.

Another miracle which has a direct bearing upon the events of the last week is the Feeding of the Five Thousand. Here Jesus is looking towards the Last Supper when during the course of the meal, he took the bread, blessed it, broke it and gave it to the disciples. The same procedure is adopted at the Feeding: 'and looking up to heaven, he blessed, and break the loaves; and he gave to the disciples to set before them.' In the early church the act of 'breaking the bread' was a hall-mark of a Christian and the fact of sharing the food would suggest a communion. Certain overtones are missed in this account for the modern world which would have been plain to a Jew. The Hebrews eagerly awaited the last days when God and man would be united seated around a table enjoying a common meal. Jesus himself makes use of this tradition in some of his parables about feasts and banquets (see Lk 13[29], Matt 8[11], Lk 14[16-24], Matt 22[1-10]). Now that the Christ had arrived on earth, the messianic feast can take place. Once more the evangelists can return to their favourite theme that Jesus fulfilled the law and the prophets. As God provided manna in the wilderness for Moses and his followers (Ex 16[14]) and food for Elijah at the brook Cherith during the drought (1 Kings 17[1-7]), here again food is being provided for God's people.

The feeding of the five thousand should not be taken at a superficial and literal level. Jesus' ability to feed a multitude upon so small an amount of food was not the important aspect of this miracle. Not only had Jesus refused earlier to turn the stones into bread with the statement, 'Man shall not live by bread alone but by every word

MIRACLES 59

that proceedeth out of the mouth of God', but there is the difficulty of the second feeding of the four thousand in Mark 8[1] and Matthew 15[32]. Had the disciples been aware of Jesus' ability to produce food at will, their question, 'Whence shall we be able to fill these men with bread here in a desert place?' would appear unnecessary. Many writers have attempted to overcome this supposed difficulty by saying that the evangelists obtained two accounts of the same event, with the details slightly different so that they confused the issue by making two incidents out of one event. On purely rational grounds this is a doubtful statement to make because if this had been so, the evangelists must have been aware that they were making Jesus perform a less miraculous miracle the second time. To feed five thousand men there were five loaves and two fishes and twelve basketsful left over; to feed four thousand men there were seven loaves and a few small fishes but only seven basketsful remained!

By far the greatest obstacle to an acceptance of the miraculous in the gospels for modern man is the breaking or abrogation of the 'normal' pattern of nature. To many, because miracles appear as irrational events, they are dismissed as 'fairy tales' or 'myths'. Yet the reason for the belief that they are irrational is that they do not conform to man's own limited conception of cause and effect. However, when we are dealing with the miraculous in the gospels, we are concerned with the prime cause of all, namely God. Biblically there is every reason to accept that there is an intimate connection between God's creation and his care for it. Jesus himself expressed this in the passage on the Sermon on the Mount commencing, 'Consider the lilies of the field. . . .' God creates by his word, as in Genesis: 'And God said . . .' let this or that happen, and it did. Again, there is the passage in Isaiah 55, 'And my word shall not return unto me void but shall accomplish that whereunto I send it.' Herein is contained the faithfulness of God to maintain the harmony and balance of nature. In so far as God's anointed was with man, a special care was being exercised over creation.

Seen from this point of view such miracles as the Stilling of the Storm or the Walking on the Water have a purpose. They reiterate that fundamental Biblical doctrine of God the creator. This does not mean that all questioning concerning the miracles is silenced. On the contrary, fresh and more pertinent questions present themselves. The context into which these miracles are placed by the evangelists shed further light on their significance for them. The details that are emphasized bring out the essential teaching of the miracle. Moreover, the symbolism and imagery of these occasions must be studied against Old Testament allusions to be fully appreciated. All of this is not avoiding the issue, but facing up to the fact that God is creator, and proceeding from there along intelligent lines to attempt an understanding of his will. A much greater act of blind faith is required by those who try to rationalize the miracles. For example, that Jesus knew of certain sand-banks when he walked upon the water, or that Jesus knew the reason for the sudden cessation of storms on the Lake of Galilee because of certain geographical and scientific facts, which we too now know. A rationalist would infer that it is coincidental that the storm ended just as Christ said, 'Peace, be still.'

Certain events have in the past been claimed as miraculous which are open, upon consideration, to a natural explanation. The result of this has been to lead commentators to assume that anything that is not immediately intelligible must be miraculous. Further, some of the 'supposed' miracles show a vindictive and capricious spirit that is not in keeping with our understanding of Jesus' character.

MIRACLES

Two obvious examples of this are the Cursing of the Fig Tree and the Coin in the Fish's Mouth (Matt 17$^{24-27}$). The parable of the Fig Tree will be discussed in its context during the events of the Last Week. However, the attempt to make the Coin in the Fish's Mouth into a miracle is the result of a too solemn view of Jesus' character. He must have had some sense of humour. When Peter was worried about paying the taxes, Jesus merely informed him that he was a fisherman and a few hours' work would be sufficient to pay for both of them. To make it any more than a humorous rebuke not to worry Jesus with mundane affairs, violates the whole tone of the passage.

POINTS FOR DISCUSSION

1. For what reasons do you think the evangelists included these miracles in their accounts? What do they tell us about Jesus?
    (*a*) Feeding of Five Thousand.
    (*b*) Stilling of the storm.
    (*c*) Raising of Jairus' daughter.
    (*d*) Curing blind Bartimaeus.
2. State the reason given in the gospels for Jesus:
    (*a*) Healing the paralytic man.
    (*b*) Not performing any miracles at Nazareth.
    (*c*) Refusing to perform miracles on other occasions.
    (*d*) Criticizing people for their attitude towards his miracles.
    (*e*) Refusing to give a sign to the people.
3. What reply did Jesus give to John the Baptist's disciples who asked, 'Art thou he that cometh, or look we for another?' With evidence from the gospels explain what you think Jesus meant by his reply.
4. Give a careful account of the miracle which Jesus performed after the Transfiguration and of the teaching he gave to the disciples at that time. (A.E.B., 1958.)

8

PARABLES

A. *The form of the parable*

MUCH of Jesus' teaching was cast in the form of parables. Today this form of teaching is rarely used so that, when reading the gospels, one is aware of something rather old-fashioned which possibly has little bearing upon the present situation. Perhaps one reason for this feeling is the fundamental difference between parables and modern thought forms. C. H. Dodd (*Parables of the Kingdom*) makes this pertinent statement, 'They (parables) are the natural expression of a mind that sees truth in concrete pictures rather than conceives it in abstractions.' Even in modern speech certain elements of concrete picture thinking are retained. A thought 'strikes' us, or we are 'frozen' to the spot in moments of fear, or a young man 'sows his wild oats'. This is a use of metaphor. Parables are an extended use of metaphor but they are also much more. Whereas metaphorical language tends to startle by a forced and striking contrast, parabolic language arises naturally from common experience. Jesus would not have considered that he was searching nature for parallels to teach certain truths. Rather he is acknowledging the fact that the Kingship of God 'is intrinsically *like* the processes of nature and of the daily life of man'. Both share a common feature in that they express a part of God's care for his creation.

Parables were never intended to convey absolute truths

covering a wide range of human activity. In many respects they are isolated truths related to specific occasions. The 'truth' of a parable is not directly stated in the parable itself but emerges when a judgement is made upon the statements. Most parables invite a judgement and Jesus himself asks men to be critical at the end of a parable. 'Which of these three, thinkest thou proved neighbour unto him that fell among the robbers?' (Lk 10[36]). 'What therefore will the Lord of the vineyard do?' (Mk 12[9]). Or the actual judgement becomes apparent in the course of the parable, 'Is it not lawful for me to do what I will with mine own? Or is thine eye evil, because I am good?' (Matt 20[15]). 'Shouldest not thou also have had mercy on thy fellow-servant, even as I had mercy on thee?' (Matt 18[33]). The thought of judgement naturally leads to a consideration of the difficulty of understanding the meaning of parables. This is a vexed question because men tend to hold emotional views about certain parables which have been established almost unconsciously over many years of familiar contact. Therefore to avoid any emotional overtones and not to run the risk of offending any religious susceptibilities let us invent a modern parable and consider its interpretation.

Parable 1

'A certain boy was playing with his ball in a park near a busy main road. Suddenly, the ball rolls off the grass into the road. The boy looks in the direction of the ball and appears to be about to dash into the road after it. However, a lady passing by says, "Stop! This is like the mouse who was playing in its hole in the wall. Suddenly smelling cheese, he rushed out to eat it. But the cat was waiting and ate the mouse. Are you going to behave like that mouse?"'

Of course this reads very strangely. Nowadays the woman would not talk in parable form but would come to the point straightway with, 'Remember your kerb drill. It is dangerous to cross a road without looking both ways and making certain that nothing is coming along.' Taken at a literal level, within the given context, this 'mouse parable' is a plea for safety first on the road.

Some people would like to go further and see here the possibility of an allegory. They would then proceed to make a list as follows, mouse = boy, cheese = ball, cat = a lorry, lady = conscience of boy. It would then be possible to elaborate the details of the story and give some interesting variants.

Other people would remove this story from its setting completely and make it a symbolic struggle of man against his environment. As the mouse is tempted by the cheese, man is tempted by his desires. Yet over and beyond the man (mouse) is the ultimate judgement (cat) upon such activities. Man must be made to recognize this.

All of these methods have been used by the evangelists in reporting and commenting upon Jesus' parables. There is no mistaking the particular reference to a specific situation. The question 'Who is my neighbour?' prompts Jesus to tell the story of the Good Samaritan. An individual might even request a particular parable, 'And when one of them that sat at meal with him heard these things, he said unto him, "Blessed is he that shall eat bread in the Kingdom of God"' (Lk 14[18]). The Pharisees were angered by the Parable of the Wicked Husbandmen 'for they perceived that he spoke this parable against them' (Lk 20[18]). Under these circumstances, where the situation is perfectly clear, it is much easier to appreciate the significance of the parable.

An allegorical approach is not missing from the gospels. Mark employs this method to comment on the parable of the Sower. Unfortunately he misses the point of the parable that Jesus told. Whereas Jesus' parable in Mark 4¹⁻⁸ is concerned with the type of soil upon which the seed fell, Mark's own interpretation is much more concerned with the ability of the seed to produce fruit under a more elaborate set of conditions than those originally envisaged. How much of this is Mark's understanding of what Jesus said is difficult to fathom. The use of the allegorical method permits each individual reader to arrive at a different interpretation.

Symbolic references and interpretations also occur in the gospels. The parables of Dives and Lazarus and of the Sheep and the Goats immediately come to mind. References to 'Abraham's bosom', and the calling of sheep and goats before the judgement seat on the last day strike the modern ear as archaic and a little unreal. Before one can appreciate these parables to any extent, one has to attempt to understand the writer's own intellectual background.

Luke is particularly concerned with the abject poverty in certain sections of society. Poverty, to him, was an evil. The Messiah would remedy the situation: 'The hungry he hath filled with good things; And the rich he hath sent empty away' (1⁵³). He reports that poor shepherds were the first to worship Jesus in the stable. There is no mention of the visit of the wise men in his gospel. When the Great Feast is prepared 'the poor, the maimed, the lame and the blind' are invited (14¹³). It is not surprising that he alone records the parable of Dives and Lazarus (16¹⁹ff.).

Matthew was a Jew who was steeped in the scriptures. His parable of the Sheep and the Goats (25³¹ff.) reflects a knowledge of such writers as Zechariah (especially 10³ff.) and Joel.

Parable 2

'Look before you leap.'

It may appear strange to call this statement a parable but there are many such parables in the gospels. 'Beware the leaven of the pharisees.' 'To him that hath shall be given, to him that hath not, even the little he hath, shall be taken away and given to another.' A better word to describe such would be proverb or adage. They derive from the Hebrew 'mashal' or wise saying concerning life. These too must be approached with caution because they are not rules or statements of universal application. As far as possible they should be considered within the context in which they occur. 'Look before you leap' may appear as a good general rule of caution but it is contradicted by an equally good proverb, 'He who hesitates is lost.' Seemingly contradictory statements are made in the parables if they are withdrawn from their context.

Fortunately Jesus himself gives a clue to the understanding of parables. When his disciples were baffled by the parable of the Sower he makes the following comment,

> 'Unto you is given the mystery of the Kingdom of God: but unto them that are without, all things are done in parables: that seeing they may see, and not perceive; and hearing they may hear, and not understand; lest haply they should turn again, and it should be forgiven them.'

The divine mystery was that God was now exercising his authority in a special manner through Jesus of Nazareth. It was possible for all men to understand it; it was not some secret knowledge that was revealed only to the disciples. They had the advantage over 'outsiders' because, by living in close contact with Jesus, they were able to see the results of this divine activity more easily. 'Outsiders' were told about the kingship of God in parables. These were extended similes which demonstrated a particular aspect of God at work. Each parable presents a challenge. The listener has to make a considerable effort to understand the truth behind the challenge. This is what Jesus meant by saying, 'Who hath ears to hear, let him hear' (Mk 4:9). Even after this effort the 'outsider' is left with part of the truth; it is only completely expressed in the life of Jesus.

Jesus is *not* saying that he teaches by parables so that people will not understand his meaning. The passage from Isaiah's inaugural vision in Isaiah 6 looks in the Revised Version as if this is the meaning. The New English Bible reading is much clearer, 'You will hear and hear, but never understand; you will look and look, but never see. For this people has grown gross at heart; their ears are dull, and their eyes are closed. Otherwise the eyes might see, their ears hear, and their heart understand, then they might turn again, and I would heal them' (Matt 13:14, 15). The complaint then is not that men cannot ever understand parables but that their present lack of spiritual insight and general stubbornness prevent them from understanding what they hear. People usually like to hear pleasant things about themselves. Jesus' parables are hard-hitting and present a picture of man which could hardly be described as pleasant. They tend to shock us when we first try to understand them but a common tendency is to sentimentalize them or to ignore the unpleasantness by looking for a less distasteful meaning.

B. *Individual Parables*

THE GOOD SAMARITAN, *Lk 10:29-37*

The Good Samaritan is Jesus' reply to the question, 'Who is my neighbour?' This is no idle playing with words but a question of fundamental importance. Jesus had already taught that we should love our neighbour as ourselves. Therefore it is imperative we should know who that neighbour is.

What disturbs us in this parable is that two men failed to go to the assistance of a wounded countryman. So often our attention is concentrated upon the commendable attitude and behaviour of the Samaritan. One tends to associate oneself with the Samaritan and experience a cosy feeling of a job well done. But the people who should really engage our attention are the priest and the Levite.

What sort of a man is it who can leave his fellow creature wounded at the side of the road? One does not need to be a Christian, or religious in any sense of the word, to appreciate the demands of such a situation. Imagine, for example, driving along a deserted country road and seeing a victim of a hit-and-run driver lying injured on the verge. The point does not need labouring; it is obvious what you would do. But the two men in the parable ignored a fellow countryman in a similar plight. One was a priest, the other a Levite; both were acknowledged as religious men! Jesus is not speculating about a highly improbable situation. There is a reason for their negligence.

The cause of their lack of co-operation is to be found in the location of the incident. It happened on the Jerusalem–Jericho road. The significance of this fact is obscured by the Revised Version's reading, 'by chance a certain priest was going down that way.' It may have been chance that led to his meeting with the injured man, but he was not travelling on that particular road by chance. Obviously both men were on their way to Jerusalem to participate in some Temple service. Had they even touched the blood of the wounded man they would have made themselves 'unclean' and therefore unfit to worship in the Temple until such time as they had purified themselves. The reason therefore for their neglect of the wounded man was their conviction that the demands that God made upon them for a pure worship were greater than the demands of the wretched individual by the side of the road. From this point of view their action was perfectly justifiable. They had failed to see that by a real concern for their fellow men they were 'worshipping' God.

Jesus means us to understand that we must not allow any cherished 'religious' convictions to blind us to the real needs of our fellow men. These demands are not made upon us because we have a vague common sympathy with all men. They are made by God because He demands respect for His whole creation, and in particular a special part of His creation, man.

It is at this point that the Samaritan assumes a significance. He had every reason to leave the wounded Jew to die. Jews and Samaritans were, for historical reasons, avowed enemies. He had to overcome these 'natural' reservations about the Jews before he could go to the man's assistance. This he did—and Jesus makes the same demands upon us, 'Go and do thou likewise.'

THE PRODIGAL SON, *Lk* 15^{11-32}

The parable of the Prodigal Son is concerned with the generosity of divine forgiveness. Perhaps it was partly because St. Paul stressed that God justifies man through faith in Jesus Christ, partly because theologians have always emphasized the sinfulness of man and his need of forgiveness, that this parable has always held a central place in Christian sentiment. It is a comforting thought that the father did not wait to receive a formal declaration of his guilt from the younger son, 'But while he was yet afar off . . . ran, and fell on his neck, and kissed him.' Then there was the rejoicing for 'it was meet to make merry'.

In all of this, one can identify oneself with the younger

son. However, it is not so happy an identification with him at an earlier stage of his career. Are most men wayward and ne'er-do-wells who are completely irresponsible in their use of money? The more accurate picture of the mass of mankind are those who lead rather uneventful lives and try to the best of their ability to do the will of God. They could say with the elder son, 'Lord, these many years do I serve thee, and I never transgressed a commandment of thine: and yet thou never gavest me a kid, that I might make merry with my friends.'

The elder son has a point of view, and a very reasonable one. It was not fair of the father to lavish his affection on the young black sheep of the family—especially if the same consideration was not shown to his brother. This attitude is contained in the statement 'I have *never* transgressed a commandment of thine,' It is smug, self-centred and self-righteous but very human. Above all things, the elder brother was fair, and in this he is a caricature of many of us. But he lacked forgiveness, the quality that the father had in abundance.

The particularly generous words in this parable are those of the father to the elder son, 'Son, thou art ever with me, and all that is mine is thine.' In so many ways he was unworthy of this statement. At least the younger brother had faced up to what he really was. He did not merely go to his father and say, 'I am sorry I have been such a source of unhappiness to you', but 'Father, I have sinned against heaven, and in thy sight.' His first concern was with his misdeeds against God and then those against his father. Just as a real concern for our neighbour is showing love for God, then a lack of love for our neighbour is a sin against God. The younger son had really learned something from his experiences and 'it was meet to make merry.'

The parable of the Prodigal Son was not intended to tell us that it does not matter if we sin occasionally, provided we know what we have done and are sorry; but it is a warning against a smugness, self-righteousness that puts fairness before forgiveness. It is a very disturbing parable.

THE UNJUST STEWARD, *Lk 16$^{1-12}$*

From what has been said about the Prodigal Son we can see an obvious connection between the elder brother and the typical Pharisee. Where confusion is caused by removing a parable from its context can be seen in the parable of the Unjust Steward. This is recounted immediately following the Prodigal Son in Luke (16$^{1-12}$). The stewards of Israel's heritage are the Pharisees and they have failed abysmally in their task. Any wide awake steward finding himself in such a predicament would, as in the parable, make the most of a bad job and use his initiative to minimize the hazards of unemployment. The Pharisees were oblivious of their impending rejection so that Jesus had to warn them to look more closely at their worldly assets to insure that there was something of value there to assist them when the ultimate rejection came. 'Make to yourselves friends by means of the mammon of unrighteousness; that when it shall fail, they may receive you into the eternal tabernacles.' If they cannot undertake the simpler task of stewardship now, then could they expect to hold any assets on their own account? If the parable of the Unjust Steward is removed from its context, then one has difficulty in not interpreting the parable to mean that Jesus is commending fraudulent behaviour.

DIVES AND LAZARUS, *Lk 16$^{19-31}$*

Greater difficulties are presented by the actual form of the parable of Dives and Lazarus (Lk 16$^{19-31}$). Folk legend lies at the root of the story, with the glorification of the poor man in heaven and the discomfort of the rich man. The details of the life-to-come are crude and bear many resemblances to contemporary rabbinic writings. Lazarus 'was carried away by the angels into Abraham's bosom' and the picture of the distinction between heaven and the other place is grim, 'between us and you there is a great gulf fixed, that they which would pass from hence to you may not be able, and that none may cross over from thence to us.' But these are the mere trappings of the parable because two major points are raised in the last section. First, there is sufficient scriptural evidence for knowing how one should lead one's life. The readings they hear in the synagogue from the scriptures should make the rich man and his five brethren know and accept their social responsibilities. 'They have Moses and the prophets; let them hear them.' When Dives persists in being allowed to return to earth to inform his brothers of their duty, a cynical remark is made about the resurrection of the dead, 'If they hear not Moses and the prophets, neither will they be persuaded, if one rise from the dead.' Luke is here commenting upon the refusal of the Jews of his day to accept the fact of the resurrection of Jesus.

PARABLES

POINTS FOR DISCUSSION

1. Look up and read the following figurative expressions in their context. Write out their meaning in simple modern English.
    (*a*) By their fruits ye shall know them. Do men gather grapes of thorns, or figs of thistles? (Matt 7^{16}.)
    (*b*) Can the sons of the bride-chamber fast, while the bridegroom is with them? (Mark 2^{19}.)
    (*c*) Everyone therefore which heareth these words of mine, and doeth them, shall be likened unto a wise man, which built his house upon the rock. (Matt 7^{24}.)
    (*d*) No man, having put his hand to the plough, and looking back, is fit for the Kingdom of God. (Luke 9^{62}.)
2. What question arises in the telling of the following parables? Write out the verse where possible.
    (*a*) Good Samaritan. (Luke 10^{25-37}.)
    (*b*) Labourers in the Vineyard. (Matt 20^{1-16}.)
    (*c*) Ten Virgins. (Matt 25^{1-13}.)
    (*d*) Unmerciful Servant. (Matt 18^{21-35}.)
    (*e*) Wicked Husbandmen. (Matt 21^{33-46}, Mark 12^{1-12}, Luke 20^{9-19}.)
3. Which parables make the following points:
    (*a*) Be prepared at all times for God's rule.
    (*b*) Persist in your prayers.
    (*c*) You must forgive your neighbour.
    (*d*) You serve God by serving your neighbour.
    (*e*) Jesus came to help the sinner and social outcast?
4. The parable of the Sower appears to refer to the reception of Jesus' teaching by the people. Discuss this, with reference both to the parable and also to Jesus' explanation to the disciples. (London, 1961.)
5. Rewrite in modern English the parable of the Wicked Husbandmen. What meaning did Jesus intend to convey by this parable? Why did the chief priests and scribes perceive that it was spoken against them? (S.U.J.B., 1958.)

9

THE KINGDOM OF GOD

The central theme of the gospel is the proclamation of the Kingdom of God. Without a sufficient understanding of this term, the gospel itself becomes meaningless. One major obstacle, right at the outset, is that Matthew calls it the Kingdom of Heaven. This is because he is a Jew and, wanting to avoid the use of God's name in common speech and the possibility of breaking the third commandment, he prefers this synonym. Yet, in doing so, he implies, to the modern reader, a place far removed from earth which is only realized at the end of life. In everyday language heaven is the opposite of earth, usually the 'place' associated with those who have died. Moreover, if we say, 'You can do that until Kingdom come', we mean that however long one performs a certain action, it will never be completed. Now both of these ideas, remoteness and incompleteness, are the opposite of what the evangelists wished to convey to their readers by the Kingdom of God.

The Kingdom of God is not a place. 'Rule', 'reign' or 'kingship' of God is a better translation of the Greek word. Therefore, we are not concerned with territorial areas or boundaries but with human relationships, how men react to the control and care of God's providence. Of course there is nothing new to the Hebrew in the idea of God as creator, controller and upholder of nature, nor for that matter in the concept of God as King (Ps 10[16], 24[7ff.]). We have already noted the important position of the kingly figure in the enthronement psalms and their particular significance for the messianic ideal. During the exile the view of God as King of the whole world had found characteristic expression in the writings of Second Isaiah. But the idea of the dominion of God naturally raised a problem seen in the book of Job, namely, if God is in control, how are we to explain the success of the evildoer? This problem had particular importance for the Jews since Yahweh was *their* God yet they were continually a downtrodden nation. They thus evolved the idea that God had laid aside a part of His power, which, at a later date, He would take up again. This idea of a future day of the Lord is very prominent in later Jewish writings, and the assumption is that on that day God will lead the Jews in triumph over all their foes.

When Jesus came into Capernaum he preached, 'The time is fulfilled, and the Kingdom of God is at hand; repent ye, and believe in the gospel' (Mk 1[15]). God had at that time taken up again His power, desiring to establish a new relationship with men through His Son, then and there. God's purpose was being revealed openly through Jesus. This involved men in true repentance and a realization of the good news that there was the possibility of harmony between man and God. Jesus proclaims this as a present fact; it has nothing to do with future realization. God's rule had come; men had to acknowledge this.

As the person of Jesus and the fact of the Kingdom of God are intimately connected, then Jesus' 'person' throws considerable light upon the nature of the Kingdom of God. Up to that time, many in Israel had believed that, by a punctilious attention to the demands of the Law, they could help to bring the Day of the Lord into being. Jesus claimed a loyalty from men which superseded the Law. Moreover, men could not assist in establishing the Kingdom of God—it was already an established fact, a reality.

'The law and the prophets were until John: from that time the Kingdom of God is preached, and every man entereth violently into it' (Lk 16[16]). The arrival of Jesus brought a crisis in human affairs. All that was evil or hindered God's purpose was judged and condemned. The choice was clear. Man could either sin or repent. 'The men of Nineveh shall stand up in the judgement with this generation, and shall condemn it: for they repented at the preaching of Jonah: and behold, a greater than Jonah is here' (Lk 11[32]). The 'greater than Jonah' demanded recognition and the 'judgement' that he brought was inescapable. It is impossible to avoid a moral or intellectual issue merely by ignoring it. By the same token it was impossible to ignore the Christ. 'He that rejecteth me rejecteth him that sent me' (Lk 10[16]). A rejection of Christ involves a rejection of God, and the consequent judgement.

Here we have arrived at the heart of the concept of the Kingdom of God, in that Jesus' ministry was a manifestation of the Kingdom. 'But if I by the finger of God cast out devils, then is the Kingdom of God come upon you' (Lk 11[20]). All his work, his miracles, his teaching and his actions bear witness to this fact. God was acting openly in the affairs of men on their behalf. But Jesus gave this idea a new and startling content. Whereas the Hebrews had considered the Day of the Lord as one of final vindication and glorification of the Jewish race Jesus showed that it necessitated suffering and humiliation. Jesus' passion and crucifixion are an example of his obedience to God's command. The sin of man, the greatest evil that man could inflict upon Christ, was endured so that, as Paul says, the self-exhibitionism of sin might be seen to be 'exceeding sinful' by comparison with the love of God! Here is the paradox, the 'mystery' of the Kingdom of God, that it was revealed in the suffering and death of the Christ.

The same love of God for man that is seen on the cross

is also present throughout the ministry. The proclamation of the Kingdom was not a 'take-it-or-leave-it' affair. It involved Jesus in positive action of seeking out the sinners, of being deeply concerned at the fate of the lost, of demonstrating God's fatherly care for his creation, and of demanding a child-like trust from man. On the negative side, it involved a rejection and condemnation of all that was evil. The fight was on,

> 'And the Seventy returned with joy, saying, "Lord, even the devils are subject unto us in thy name". And he said unto them, "I beheld Satan fallen as lightning from heaven. Behold I have given you authority to tread upon serpents and scorpions, and over all the powers of the enemy: and nothing shall in any wise hurt you. Howbeit, in this rejoice not, that the spirits are subject unto you: but rejoice that your names are written in heaven"' (Lk 10^{7-20}).

What this implies is not that men were helping to establish the Kingdom of God (that was God's gift), but were participating with Christ within the Kingdom. Christ is often spoken of as doing something for men. A more comprehensive view is to see Christ doing something with men.

From what has been said it is obvious that the Kingdom of God involves men in action. This activity can be gauged by our attitude towards Jesus himself, that is, our commitment to Christ. This commitment cannot be regarded as a single isolated fact of life. It is not like putting a rubber stamp on an article so that it is certified. The Kingdom of God is a challenge that involves man in a continual struggle right in the centre of the affairs of men. There is nothing of the bargain, or *quid pro quo* about the Kingdom, 'For many are called but few chosen' (Matt 22^{14}). 'Not everyone that saith unto me, "Lord, Lord", shall enter into the Kingdom of Heaven; but he that doeth the will of my Father which is in heaven' (Matt 7^{21}). Doing the will

A chart illustrating three important themes about the Kingdom of God

Name of parable	Reference	Important verses	Meaning
A. Kingdom of God is compared with a harvest which is ready now.	Matt 9[37, 38] = Lk 10[2]	'The harvest truly is plenteous, but the labourers are few. Pray ye therefore the Lord of the harvest, that he send forth labourers into his harvest.'	The harvest is ready now. The Kingdom has arrived.
i. Tares.	Matt 13[24–30]	'Let both grow together until the harvest.'	i. Everyone is not prepared, but the harvest cannot be delayed because there are weeds among the crop.
ii. Sower.	Mk 4[3] = Lk 8[4] = Matt 13[3]	'and yielded fruit, some a hundredfold, some sixty, some thirty.'	ii. In spite of numerous difficulties the farmer ultimately harvests his crop.
iii. Seed growing secretly.	Mk 4[26–29]	'But when the fruit is ripe, straightway he putteth forth the sickle, because the harvest is come.'	iii. The presence of the Kingdom reveals itself in a sudden and dramatic manner.
iv. Leaven.	Matt 13[33] = Lk 13[20, 21]		
B. The Kingdom is for all men.			
i. Mustard seed.	Lk 13[18] = Mk 4[30]	'the birds of the air roosted in its branches.'	i. As a tree sheltering birds in its branches is a common symbol of an empire protecting its subjects (*cp.* Dan 14[2], Ezek 17[22]); therefore all nations can enter into the blessings of the Kingdom.

ii. Drag net.	Matt 13^{47}	'the kingdom of heaven is like unto a net, that was cast into the sea, and gathered of every kind.'	ii. Jesus' ministry was a wide casting of the net. There is no discrimination. The sorting out comes later.
iii. Marriage feast. Great supper.	Matt 22$^{1-14}$ = Lk 14$^{15-24}$	'Go out into the highways and hedges, and constrain them to come in, that my house may be filled.'	iii. God is offering hospitality to all men.
C. The Kingdom demands action; if not, man is judged.			
i. Faithful and unfaithful servant.	Matt 24^{45} = Lk 12^{42}	'the Lord of that servant shall come in a day when he expecteth not.'	Each parable contains a crisis with the appearance of the chief figure. If individuals are incapable of responding to the crisis, they are judged. Jesus' ministry brings a crisis into the affairs of men.
ii. Waiting servants.	Matt 13^{33} = Lk 12^{35}	'Blessed are those servants, whom the Lord when he cometh shall find watching.'	
iii. Thief at night.	Matt 24^{43} = Lk 12^{39}	'Be ye also ready: for in an hour that ye think not the Son of man cometh.'	
iv. Ten virgins.	Matt 25^{1}	'Watch therefore, for ye know not the day nor the hour.'	

of God does not imply a removal or abdication from daily life, but a fuller participation in it. Even the most trivial act in Christ's name is important. 'For whosoever shall give you a cup of water to drink, because ye are Christ's, verily I say unto you, he shall in no wise lose his reward' (Mk 9[41]). No one can claim any special rights or privileges, as all is dependent upon the loving kindness and mercy of God. Here again is another paradox. The search for one's self should end in obedience to God's will and negation of self. 'He that findeth his life shall lose it; and he that loseth his life for my sake shall find it' (Matt 10[39]).

Up to this point it has been assumed, that the Kingdom of God entered as a reality into men's lives with the ministry of Jesus. This is true but it needs a slight modification in the light of certain statements in the gospel that make it appear that the Kingdom of God will only be established fully at some time in the future. A typical statement of this is found in Mark 9[1]. 'Verily I say unto you, "There be some here of them that stand by, which shall in no wise taste of death, till they see the Kingdom of God come with power." ' Whether Jesus meant to imply that the act of recognition of the Kingdom on the part of his hearers would be in the future, in spite of its present reality, or whether he intended to contrast the present Kingdom with a later manifestation 'with power' is an open question. Many have found a later fulfilment of the Kingdom in passages like Matthew 8[11] and Luke 14[16ff]. 'Many shall come from the East and the West, and shall sit down with Abraham, and Isaac, and Jacob, in the Kingdom of heaven.' Yet here again Jesus does not question the fact of the present reality of the Kingdom but merely recognizes an extension of the Kingdom to all men. Perhaps the original form into which these sayings have been cast has influenced their interpretation. Several of them relate to the messianic banquet which the Jews believed would inaugurate the new regime on the Day of the Lord. This does not mean that all of Jesus' parables describing a banquet should be interpreted as an allusion to this future event.

By far the largest amount of evidence used to argue a case for the future advent of the Kingdom is taken from those sayings in which Jesus predicts future calamities and specifically refers to historical events connected with the destruction of the Temple in A.D. 70. Such sayings have to be treated in a similar manner to certain prophetic utterances in the Old Testament. They are not detailed predictions made by gazing into the future but considered statements arrived at by means of a deep spiritual insight into the present. Jesus states that if certain lines of activity are pursued, then the ultimate consequences are obvious. The fact that James and John did not suffer with Jesus in Jerusalem (Mk 10^{35-40}) does not mean that Jesus predicted wrongly, or that one has to search for fulfilment later in the early Church. The underlying principle was that the disciples' commitment to Christ would involve them in persecution and suffering.

B. *What the arrival of the Kingdom means for man*

The teaching of Jesus concerning the Kingdom of God is scattered throughout the gospels. A considerable amount of reading, and comparison of statements, is necessary before one can arrive at a balanced view. Several statements about the influence of the Kingdom upon man have been collected below in a convenient form. This list is not exhaustive but points in the direction of a fuller study of the Kingdom.

1. Entry to the Kingdom involves sacrifice, so count the cost

Parable of the Hid Treasure and Costly Pearl, Matthew 13^{44-46}.

Parable of the Tower Builder and King going to war, Luke 14^{28-33}.

Compare with the background of such teaching contained in Matthew 8^{19-22}, Luke 9^{57-62}.

2. The Kingdom is not a type of 'reformed Judaism'

Parable of the Patched Garment and Old Wineskins, Mark $2^{21, 22}$; the content and meaning of the Kingdom cannot be contained within the framework of Judaism.

Parable of Bridegroom, Mark 2^{18-19}. The Kingdom brings joy to men so that the mechanical observance of a ritual like penitential fasting becomes meaningless.

Compare with the teaching contained in Mark 10^{17-22}, Mark 7^{1-23}, Mark 12^{41-44}.

3. The Kingdom is concerned for the individual, especially the social outcast and sinner

Three parables in Luke 15 of the 'lost'. Coin, sheep and Prodigal Son. The joy of finding that which was lost. In the Prodigal there are echoes of the contemporary situation, the elder son (self-righteous Jew) compared with younger son (repentant sinner).

Compare with the teaching contained in Matthew 21^{28-32} (Two Sons); Matthew 22^{1-13}, Luke 14^{16-24} (Great Feast).

4. Entry to the Kingdom is not on merit but by God's grace

Parable of Labourers in the Vineyard, Matthew 20^{1-16} Divine generosity is stressed and there is no quibbling about entry; 'Once you are in, you are in.'

This parable can be contrasted with Mark 2^{16-17}, Luke 7^{36-50}, and considered as an answer to the question, 'Why does Jesus associate with worthless types?'

5. The arrival of the Kingdom involves a struggle with evil

Parable of the Strong Man despoiled, Mark 3^{27}, Luke 11^{21-22}. It should be noted that the struggle begins with the advent of Christ; it is not in the future.

Compare with the teaching in Luke 10^{17-20}, Matthew 4^{1-11}, Luke 22^{35-38}.

C. What is the importance of the Kingdom today?

Partly for the sake of certainty, partly for the sake of simplicity, men have always tried to systematize or codify Christian conduct. To this, the evangelists were no exception. All three report a statement by Jesus about salt and each one treats the saying in a different manner. The Marcan version is the simplest,

> 'Salt is good: but if the salt have lost its saltness, wherewith will ye season it? Have salt in yourselves, and be at peace with one another' (9^{50}).

Here salt is regarded as a quality which the Christian community should possess because it produces peace. As it comes at the end of a section on the precedence amongst the disciples, perhaps Mark wishes to convey the meaning that Christians are hospitable through the symbol of sharing the salt. Matthew reports the saying differently,

> 'Ye are the salt of the earth: but if the salt have lost its savour, wherewith shall it be salted? It is henceforth good for nothing but to be cast out and trodden under foot of men' (5^{13}).

The point here is explicit. Christians are the salt of the earth and they provide a purifying and preservative influence within society or else they are rejected. Luke makes no comment upon the salt statement,

> 'Salt therefore is good: but if even the salt have lost its savour wherewith shall it be seasoned? It is fit neither for the land nor for the dunghill: men cast it out' (14^{34}).

The context gives us a clue to Luke's meaning. As the statement concludes a passage about the disciples meeting the heavy demands made by Christ, 'So therefore whosoever he be of you that renounceth not all that he hath, he cannot be my disciple', salt is a personal virtue of the individual Christian which strengthens him to face the necessary privations. Savourless salt is the would-be disciple who is incapable of renunciation.

Such is the variety of interpretation offered by the evangelists. Obviously the actual context, within which Jesus said these words, has been lost. One scholar has suggested that the meaning of the original remark might be found by deduction. Salt is a valuable commodity to men, essential to life but it becomes worthless if it loses that one vital quality which characterizes it. In Jesus' ministry the most flagrant example of a failure to live up to its possibility was Judaism itself. Therefore the original saying may have been a comment upon the contemporary religious situation amongst the Jews. The development of Hebrew history suggested the possibility of the acceptance of the Christ by the Jews, but they had failed at the last step—they had become savourless salt.

None of this implies that the evangelists were wrong or not telling the truth. Total rejection of this kind is like emptying the baby with the bath water. Man's understanding of the Christ or the Kingdom of God is dependent upon a variety of factors. As Christianity is a revelation set within an historical framework, it necessarily involves a study and understanding of the documents relating to those historical events. However, that understanding will be conditioned by other factors such as the dominant ideas current at the time in which the study is made, or by the influence exerted by the Holy Spirit. Jesus said that the Kingdom had arrived. The demands this makes upon men have been understood in many different ways.

Some men may be more right than others, but all men only express a relative and partial truth of the Kingdom. The greatest danger lies in an attempt to foist one's own subjective partial truth upon others as *the* absolute truth, the 'gospel truth'. If the study of the scriptures is confused with homiletic statements, the dangers are legion. Yet both a study of the scriptures and homiletic statements have an important part to play in the life of Christians. The difficulty is keeping them separate and not confusing one with the other.

POINTS FOR DISCUSSION

1. What do you understand by the term 'the Kingdom of God'? Illustrate your answer by reference to four parables about the Kingdom.
2. 'The Kingdom of God is a present reality. God's Kingdom will come.' Explain by reference to two parables how both views were expressed by Jesus. Do you think these statements are contradictory or can they be reconciled?
3. What proof did Jesus give his disciples that 'the Kingdom of God was at hand'?
4. Recount the parable of the Seed growing secretly (*not* the Sower), and the Mustard seed. What do these parables tell us about the kingdom of God? (S.U.J.B., 1959.)
5. 'Thou art not far from the Kingdom of God.' In Jesus' teaching which qualities are stressed for those who wish to be members of the Kingdom?

10

THE SERMON ON THE MOUNT

A. The problem of understanding the Sermon on the Mount

No single section of Jesus' teaching has caused greater controversy and heart-searchings than that material contained in Chapters 5–7 of St. Matthew's gospel, commonly called the Sermon on the Mount. One reason for this is that, although there is a considerable measure of agreement about what Jesus said, there is much room for disagreement about his original intentions in making such statements and what meaning he wished them to have.

In the first instance it is clear that Matthew intends us to see Jesus 'going up a mountain' to teach the people certain precepts about life. The whole situation is reminiscent of Mount Sinai. On this occasion it is not the voice of God stating the laws but Jesus himself who says, 'You have heard that it was said . . . but I say unto you'. In the course of the teaching Jesus strengthens this impression by direct reference to the Law and the prophets, and the traditions about the Law. But it is at this point that one begins to question Matthew's assumptions. It may be perfectly easy in a simple pastoral society to make certain laws, or rules of conduct, which decent and honest men can observe, particularly if the adherence to such rules have been verified by men's actions and they relate to commonly accepted standards such as not committing murder, not stealing, not lying in a court of law, or not breaking up families by committing adultery. The situation is totally different when one is attempting to formulate rules governing a highly complex society in which such a variety and diversity of circumstances influence personal relationships. This is especially true if one tries to control, not the outward acts of men, which are apparent to all, but the motive behind the act, which is known only to the individual. For example, society can deal quite effectively with the murderer but it is not competent to prevent men from having hateful or revengeful thoughts. Jesus' teaching in the Sermon states clearly that vindictive and hateful thoughts are a species of murder!

Certain scholars were not perturbed by such difficulties. Their argument was that no one should consider the Kingdom of God in terms of the existing state of society. Jesus was talking of the perfect society, giving instructions which have divine approval, to men who have been recreated and reborn through their experience of entering the Kingdom of God. Anything less than complete acceptance of the literal meaning of Jesus' statements would not do justice to the situation. 'Obey these instructions and all will be well.' What is a little disturbing in this perfectionist attitude is that the demands that it makes upon man are so great as to appear impossible to fulfil. A great danger lies in the fact that many people might be inclined to reject Jesus if they thought he demanded a standard of conduct which it was impossible for them to achieve. The perfectionist is quite correct in affirming the seriousness of Jesus' pronouncements but unrealistic in the way he would wish to enforce them. Another objection to the perfectionist is that the Sermon on the Mount is only a small part of Jesus' work and teaching and that over-emphasis upon any particular aspect of his preaching outside the wider context of the ministry leads to a grotesque distortion, even to an untruth. A pertinent question to ask here is, 'How far is it possible for a complete and absolute truth to be stated in verbal form?' Words are notoriously difficult to handle, mainly because they mean different things to different people. Normally this does not matter in everyday life because only a relative measure of agreement about meaning is needed. But to claim, as the perfectionists do, that absolute meanings and truths can be stated is difficult to accept. If this were possible, a large number of the legal profession would be unemployed at one stroke!

Another way of attempting to get at the truth of the Sermon on the Mount starts from the point at which the perfectionist attitude falls down. Man is incapable of fulfilling Jesus' demands and this makes them worthy of our serious attention. It is only by being made aware of his own limitations and shortcomings that man can throw himself wholeheartedly upon the mercy and loving kindness of God. The Law is the means by which man learns of his imperfections. Once man recognizes his own sad plight, he is then willing to accept the gospel seriously. The Sermon on the Mount, as part of the divine law, is teaching man his own inadequacy so that he is in a position to receive the grace of God. All of this sounds reasonable and is particularly acceptable to those who have a knowledge of what Paul said about the law being a 'teacher' or 'tutor' of men (Gal $3^{23ff.}$; Rom $7^{7ff.}$). Such a conclusion is the result of reading too much into the Sermon on the Mount. Nowhere in the Sermon does Jesus reflect upon the inability of men to do God's will; rather the reverse is true. Jesus expects his disciples to obey his commands. 'Everyone therefore which heareth these words of mine, and doeth them, shall be likened unto a wise man . . .' (7^{24}). In the concluding section four pictures are given of the narrow and the wide gate, of the sound and the bad trees, of men standing before the throne of God at the Final Judgement and of the building of a house on rock and sand. All of these stress the importance of man's own will in striving to achieve his salvation. Therefore reasonable as the imperfectionist argument sounds, it will not fit the facts of the case.

A third and final attempt to understand the Sermon has been suggested, which eliminates many of the difficulties of the first two theories. The demands of the Sermon are impossible to meet because they were intended to meet special circumstances of the situation in which men found themselves at the time of Jesus. Christ had come to give men a last chance of repentance; the end was near. The Sermon represents a type of 'Martial Law' for those living in a time of crises. Extreme measures were needed to meet an extreme situation. Love your enemies, reject normal methods of retaliation ($5^{38\text{ff.}}$), 'and if thy right hand causeth thee to stumble, cut it off, and cast it from thee' (5^{30}). Although there is an element, and an important element, of the truth in such a view, namely, that Jesus was always aware of the crisis that he had introduced into the affairs of men, it is all too one-sided. Jesus is equally conscious of the certainty of salvation that God offers, a certainty that is valid for all men at all times. If this is not so, then the Sermon merely has an historical interest in being the inspired utterance of a religious fanatic. This Jesus was not, but he came to give life to ordinary men and women at all times.

None of the previous arguments solves the problem of the Sermon, yet they all add to an understanding of its nature. As diverse as these solutions are, they all share two common presuppositions; first, that the Sermon is a complete and sufficient statement of the facts. Second, that the Sermon is a statement of law. Both of these ideas need challenging if further progress is to be made in discovering the truth.

B. Certain presuppositions reconsidered

1. *That the Sermon on the Mount represents a complete and sufficient statement by Jesus regulating the conduct of men.*

The Sermon on the Mount undoubtedly contains many original sayings made by Jesus. Whether they were all made at one time, and in the order in which we now have them, there is considerable doubt. One fact which substantiates this conclusion is that there is another collection of the sayings of Jesus in Luke 6^{17-49}, known as the Sermon on the Plain, sharing much common material with Matthew's collection, but varying considerably in emphasis. For example, both passages deal with prayer and their content can be summarized as follows:

Matthew 6^{6-15}	Luke 11^{1-13}
1. Do not be like hypocrites and make a show of praying.	1. Teach us to pray.
2. Do not use empty phrases.	2. Lord's Prayer.
3. Take the Lord's Prayer as a model.	3. Persist in prayer. (Friend at midnight.)
4. When you pray, also forgive.	4. Ask and God will give you.
	5. Father who does not fail to give his sons gifts.

Both evangelists have at their disposal certain facts about Jesus and certainly some of his statements. The form into which they mould their information will depend not only upon the type of people for whom the gospel is intended, but also upon the teaching about Jesus that has been received in the evangelists' church. Scholars have long pointed out the difference between the proclamation of certain facts about Jesus and the teaching based upon these facts. In the example quoted above, Matthew is writing for a community that is familiar with prayer, but needs guidance as to the procedure to be adopted. Luke, however, is much more concerned to encourage his readers to pray. They were not familiar with prayer, as he understood it, and he does not want them to be dismayed too easily in acquiring the habit. Without stretching the imagination too much, one is aware of Matthew's Jewish–Christian background, whilst Luke is writing with a Gentile–Christian community in mind. Consequently, there is a danger in taking either of these statements as final or absolute in themselves.

Several statements made by Jesus in the Sermon on the Mount, if isolated from the context of the gospel, do not make sense. For example, to tell Christians, 'Ye are the light of the world', is nonsense when one remembers the shortcomings of the Twelve, as they are fully reported in the gospels by the evangelists themselves (5^{14}). It only becomes a meaningful comment when one remembers Jesus' words, 'I am the light of the world', so that Christians, true to the name, should emulate their master. Jesus also says, 'Till heaven and earth pass away, one jot or one tittle shall in no wise pass away from the law' (5^{18}). He then proceeds to demonstrate the weakness of the law! What Jesus is referring to here is not the whole Mosaic law but those passages in scripture foretelling Messiah which will all be fulfilled in him.

Moreover, a very peculiar theology can be constructed from isolated passages in the Sermon on the Mount. The passage on forgiveness could, on the evidence of this passage alone, be understood in the form of a bargain between man and God. 'For if you forgive men their trespasses, your heavenly Father will also forgive you. But if you forgive not men their trespasses, neither will your Father forgive your trespasses' (6^{14}). There is no such commercial spirit in Jesus' teaching as is seen in the parable of the Unmerciful Servant, but forgiveness is a necessary corollary of God's own generous nature,

'Shouldest not thou also have had mercy on thy fellow servant, even as I had mercy on thee?' (Matt 18:34).

The sayings on the Sermon on the Mount are not complete and sufficient within themselves, but must always be considered within the total framework of the gospel and with particular reference for the community to which they are addressed.

2. *'The Sermon on the Mount is law'*

Before commenting on this sentence, it would be as well to consider the characteristics of law. The law is the means by which men control relationships within society. It presupposes certain standards of conduct which, if they are broken, are ascertainable by the behaviour of a given person. Evidence can be produced and the necessary judgement made. In many respects law is a negative force—negative in the sense that it operates by the default of a citizen. Men are obliged by law to pay taxes, but the law only comes into operation when they do not pay them. By its nature law puts all the onus upon man to obey; it presupposes that man can and will perform all the necessary requirements of the legal system by his own efforts.

Much of what Jesus says in the Sermon on the Mount is concerned with regulating relationships between men. But, unlike the civil law, Jesus' commands are concerned with attitudes, and reasons or motives for actions, rather than the actions themselves. Therefore there can be little that is accurately known, by laws of evidence, to the general public. 'Love your enemies, and pray for them that persecute you' (5:44). What evidence, in a legal sense, can be given of a man's love of his enemy? Love for an enemy may involve considerable hostility and persistent pressure, for this is love in a complete sense, not the sentimental sort of 'love' that agrees with an avowed enemy and lets him have his own way. Moreover, a man might behave correctly so far as the world is concerned, but from the wrong motives. The judge here is the individual concerned; the world may never know.

Whereas the law operates in a negative sense, by its default, Jesus demands a positive attitude. Those who inherit the Kingdom of God are the poor in spirit, the peacemakers and the persecuted! It is not as simple as 'Do this', or 'Do not do that', but 'Do what you must do with the right attitude.'

A man can boast about keeping the law because he achieves success by his own efforts. No man could possibly keep all the demands made upon him by Jesus in the Sermon on the Mount at all times. When he does achieve success, it is only partly by his own efforts because a great deal of strength and comfort has been derived from God as well.

So far, much has been said about what the Sermon on the Mount is *not*. It is now time to see what it *is*. The Sermon is a collection of the sayings of Jesus, on a variety of topics, intended as part of the catechitical training of those who have recently entered the Christian Society. As part of a catechism, or rules and regulations of a Christian Society, it assumes that the person interested in the Sermon will recently have undergone a change in his life by having heard the gospel. Therefore the individual statements included in the Sermon have to be studied within the framework of the gospel itself. The convert already knows the power and force of Christ's entry into the world for his salvation. Here in the Sermon is formed the picture of the ideal member of the Kingdom of God. The person just initiated has been recreated, reborn through the power of Jesus Christ so that his life is no longer his own, but belongs to the City of God. The convert can share in the experience of establishing God's Kingdom on earth. He will have to strive much, but God's grace is there to support him all the time.

C. *An analysis of the Sermon*

5¹⁻¹²: *Beatitudes*

Matthew begins by stating the qualities that are most conspicuous in those desiring to receive the blessing (that is, the full riches) of God's loving kindness and mercy. Such men are 'poor in spirit', not being arrogant or self-assertive; 'those that mourn', men who are aware of the sin and suffering of mankind as a whole; 'the pure in heart', men with singleness of mind, without guile. In many ways the Beatitudes represent a veiled witness to Jesus as the Lord of the 'poor in spirit', 'meek' and the 'merciful', etc.

5¹³⁻¹⁶: *True discipleship (salt and light)*

A man possessing the qualities outlined in the Beatitudes is a true disciple of Jesus and, as such, he has a function to fulfil; just as salt gives savour and a light shines, a disciple bears witness to Jesus Christ.

5¹⁷⁻²⁰: *Statement of main theme. A new righteousness*

Witness to Jesus Christ involves both the realization that he fulfils all the scriptures concerning the Messiah and that he demands from man a certain type of behaviour. Such behaviour is the result of personal knowledge of Jesus Christ, that he lived, died and was resurrected for man. Conventional piety and adherence to outward forms are not enough. 'For I say unto you, that except your righteousness exceeds the righteousness of the scribes and Pharisees, ye shall in no wise enter into the Kingdom of God.'

In this pronouncement three sections of the community

THE SERMON ON THE MOUNT

are mentioned, the scribes, the Pharisees and the disciples. The scribes represent the professional theologians, the Pharisees pious layman intent upon obeying God's law, and the more recent section of community, the disciples, men who have acquired a new insight into God's will through Jesus. Matthew then groups certain sayings, contrasting the professional theologians and pious laymen with the righteousness of a true disciple.

5^{21-48}: *Professional theological arguments considered*

Matthew proceeds by taking six statements made in the law, or the traditions on the law, and giving six antitheses, or challenging statements to the contrary, made by Jesus. These antitheses cover a wide area of human conduct and refer to such matters as a proper attitude towards a brother, to women, marriage, truthful speech and behaviour with regard to an enemy.

'Ye have heard that it was said to them of old time, "Thou shalt not kill"; and whosoever shall kill shall be in danger of the judgement: but I say unto you, that everyone who is angry with his brother shall be in danger of the judgement.' Here Jesus is claiming that one who indulges in vindictive or spiteful behaviour towards a brother is as guilty as a murderer. Rationally, if one is considering motives, there is much truth here because both attitudes are rooted in hate. Unfortunately the English translation of the text of the next verse makes it appear that one would take a person before the Sanhedrin for saying, 'Fool'. What is implied by the two words translated as 'Raca' and 'fool 'is a malicious and vindictive attack upon a person for intellectual or moral depravity. Then follows two illustrations of a right attitude. The first stresses that reconciliation with your neighbour must have priority over making an offering to God in the Temple. To a Jew, Temple business took precedence over everything else, so that Jesus is claiming that reconciliation is much more important than sacrifice. Secondly, Jesus asks that one does not have a stubborn insistence upon one's own rights so that litigation becomes necessary. The forgiving spirit is preferable (here is a realistic touch) if only because it is cheaper!

'Ye have heard that it was said, "Thou shalt not commit adultery": but I say unto you that everyone that looketh on a woman to lust after her hath committed adultery with her already in his heart.' Once again Jesus goes to the heart of the law because a lustful attitude is at the root of adultery. This does not mean that one has to be excessively puritanical. Sexual desires, like any other desires, are perfectly natural and there are legitimate ways of satisfying them in marriage. What is being condemned is a lascivious nature, a dirty-minded attitude and smuttiness. It would be a poor day for a society when it is considered wrong for man and woman to work side by side and not to derive pleasure and satisfaction from the mutual contact. A much more difficult saying about divorce follows. 'I say unto you, that everyone that putteth away his wife, saving for the cause of fornication maketh her an adulteress.' Jesus forbids divorce except on the grounds of adultery. At the expense of appearing blasphemous, this strikes one, as it stands, as being unrealistic. Men and women do divorce one another and good legal grounds may be stated for doing so. Mutual incompatibility and cruelty immediately come to mind. Moreover, the Jewish divorce law was a liberal and enlightened move to protect the woman, the weaker partner, by allowing her some social security in permitting her to remarry. However, the saying must be taken within the context of the gospel and not isolated. In Mark 10^{2-12} Jesus condemns divorce altogether! But the reason he does so is instructive. Jesus does not rely upon Moses but goes back to the creation story (v. 6) and says that when God rules men's lives laws are superfluous. Under the Kingship of God, divorce will no longer be considered because then the original purpose of God, that there is a Paradisal harmony between the sexes, will be established. No one assumes this is easy but if the Kingdom of God is real, then divorce is out.

Another antithesis deals with the 'lex talionis' the right of claiming that the degree of punishment should correspond to the extent of the offence. 'An eye for an eye and a tooth for a tooth.' 'But I say unto you ... Whosoever smiteth thee on thy right cheek, turn to him the other also.' Great confusion has been caused by applying this as a general precept, governing Christian humility. Whenever Jesus talks of insult, persecution and dishonour to a disciple, it occurs because of the discipleship itself. Here also a deliberate act of violence is envisaged because of a person's discipleship. Jesus maintains that the normal retaliatory measures should not pertain here, but evidence of true discipleship is to bear the insult and forgive the injustice. This is but part of the cross we bear, if we follow Christ.

Finally, Jesus teaches his followers to love their enemies. This is a very positive demand; it requires not merely refraining from doing harm to others, but a genuine and sincere regard for others, which is based upon the same principle as self-esteem. 'Thou shalt love thy neighbour as thyself' (Matt 22[29]). If one behaves in this way then one is bearing witness to God who from His own generous nature 'maketh the sun to rise on the evil and the good, and sendeth rain on the just and the unjust'.

One word of warning is needed before closing this section. The scribes, the professional theologians of Jesus' day, were no fools. They were, for the most part, decent honest men earnestly seeking to do the will of God. Jesus' criticism of their position stems from the fact that he, the

Christ, is now present amongst men and the position has changed. Men have the possibility of forgiveness, and salvation is theirs. However, this requires a change of character and a different disposition. Perhaps it is misleading to talk about ethics and morality. Jesus did not ask his disciples to do more good works than their Jewish contemporaries or to be 'better' people morally. A disciple must be obedient to God's demands as these were revealed to him through the Christ. The scribes did not acknowledge Jesus as the Christ. Therefore they missed their greatest opportunity.

6^{1-18}: *Pharisaic religious piety*

Matthew continues by giving three examples of 'doing righteousness', that is, the performance of those obligatory religious duties, almsgiving, prayer and fasting. In all of these examples the necessity of secrecy and privacy is stressed in contrast with the blatant self-ostentation of some contemporary hypocrites. Hypocrite means, literally, an actor. The true basis for any form of worship is contact and communion with God. If one desires to obtain only the praise of men this can be achieved, but that is all. Once again Jesus stresses the importance of the motive for the act rather than the act itself.

$6^{19} - 7^{29}$: *Typical attitudes towards life by a true disciple*

The central theme of this last section is, 'But seek ye first his Kingdom, and his righteousness, and all these things shall be added unto you' (6^{33}). It is not intended as a systematic and comprehensive statement of the Christian way of life but gives certain pointers in difficult aspects of life. Men all need a sense of security, but this is less likely to be achieved by the accumulation of worldly objects, which are liable to deteriorate in the course of time, than by the confidence placed in God's providence to care for His children. Yet this does not mean that man is passively waiting for God's generosity to assert itself. It requires a positive effort upon the part of man, by witnessing before all men his genuine desire for the establishment of God's Kingdom upon earth.

POINTS FOR DISCUSSION

1. In what sense can the Sermon on the Mount be described as a practical guide to Christian behaviour?
2. 'But seek ye first his kingdom, and his righteousness; and all these things shall be added unto you.' What did Jesus teach about this in the Sermon on the Mount? State *briefly* what you think he meant.
3. 'I came not to destroy, but to fulfil.' Explain, with three illustrations from the Sermon on the Mount, how Jesus gave a deeper meaning and greater significance to the Law.
4. To what did Jesus liken his disciples in the Sermon on the Mount? What special commands did he give them about praying and fasting?
5. To which of the ten commandments did our Lord refer in the Sermon on the Mount? What was his own teaching on each? (Ox. and Camb., 1958.)

11

THE DISCIPLES OF JESUS

ALTHOUGH Jesus is the central and main character of the gospel narrative, this must not detract from the importance of the group of people who were closely associated with him during his ministry. The disciples are important if only because they are the men who assisted in establishing God's Kingdom upon earth after Jesus' death. For this reason they have been respected and revered by the church at all times because they had the privilege of knowing and working with Jesus whilst he was on earth. However, there is a danger in assuming that they are peculiar people, different from us, the saints. A gulf is created between them and us. This view distorts their true character. They were ordinary men and as such reflect the human response, our type of reaction, to the person of Christ.

The followers of Jesus were known by a variety of names. Disciple comes from the Latin word 'discipulus', a pupil, and the Greek term for disciple has the same root as our word 'mathematics', meaning someone who learns. A pupil–teacher relationship is understood by using this term. It is true that the disciples called Jesus Rabbi or Rabboni, master and teacher. As well as receiving instruction, the disciples also gave it. When out on a mission preaching the Kingdom of God, they are better known as apostles, that is, people sent out to do a specific job. Another aspect of their function is seen in the term minister or servant because Jesus taught them that their true calling was to minister to others; from the account of Peter's denial, it is possible that they were called the Nazarenes because of their close association with Jesus of Nazareth. Therefore the word disciple can be used in a wide sense covering all those people who had a close relationship with Jesus.

Perhaps the best known term to describe the disciples in the gospels is simply 'the Twelve'. Lists of the names of the Twelve can be found in Mark $3^{13ff.}$, Matthew $10^{1ff.}$ and Luke $6^{12ff.}$. A careful scrutiny of these lists will reveal that there are discrepancies in the names. Taking Mark as the norm, Luke inserts Judas the son of James and omits Thaddaeus. Mark himself also leaves out Levi, the son of Alphaeus, whose call he particularly mentions in 2^{13}. Unless a change of name was coincident with becoming a disciple, as in the case of Simon Peter, Levi the son of Alphaeus becomes James the son of Alphaeus, or Matthew the publican. Another difficulty in accepting twelve as a fixed and rigid number of disciples is that it was Cleopas, who had not been mentioned before, and another to whom Jesus made himself known on the Emmaus road by the breaking of bread on the first Easter Sunday. A way out of these difficulties is given by the passage in Acts $1^{16ff.}$, referring to the election of another apostle to take the place vacated by Judas Iscariot, where it is specifically stated that there was a choice of men for the office who had been with Jesus from his baptism until the time of his resurrection. Consequently we are to assume that Jesus was surrounded with a large number of disciples, some of whom were particularly close to him.

The evangelists all employ the term 'the Twelve' to designate the disciples. Therefore the tradition of the Twelve must have had some significance for them. A clue is found in Matthew $10^{5, 6}$ when Jesus is giving final instructions to the disciples before sending them on a mission, 'Go not into any way of the Gentiles, and enter not into any city of the Samaritans: but go rather to the lost sheep of the house of Israel.' The Twelve tribes of Israel in the Old Testament are symbolically replaced by the Twelve disciples. The new Israel found its origin in Jesus and his disciples. That this is accepted by the other evangelists is demonstrated by the events at the Last Supper. Whereas God had ratified his covenant with Israel, the twelve tribes, through the blood of a sacrifice, Jesus institutes a new covenant through his blood, with the New Israel, in the upper room. Luke, moreover, with his usual interest in the Gentile mission, finds a place for that outer ring of disciples within the original intentions of Jesus. Just as Moses, when his work expanded, needed to appoint seventy elders to instruct Israel (Num 11[16]), Jesus appointed 'seventy others' to preach the gospel (Lk 10[1ff.]).

A surprising fact about the disciples is their immediate response to a call by Jesus. ' "Come ye after me, and I will make you to become fishers of men." And straightway they left their nets and followed him' (Mk 1[17, 18]). Surely the historical accuracy of such a statement must be questioned? In the Fourth Gospel a considerable amount of preparatory teaching is given to the disciples by John the Baptist and they even discuss the matter with Jesus. But the synoptic evangelists are not interested here in the historical truth of the situation but, writing after the event, cannot see how anyone would hesitate in answering a call by the Christ. It is rather like the parable of the pearl of great value; the disciples' intuitive response to their call and their wholeheartedness and singlemindedness to become members of the Kingdom of God is an inspiration to others. Perhaps it would be as well, at this point, to dispose of the crude idea that Jesus had 'poor and ignorant' fishermen for disciples! Firstly, James and John, when they were called, left their father 'with the hired servants'. There was nothing poor about them. Secondly, John's gospel informs us that, during the trial before Caiaphas, whilst Peter remained in the outer courtyard, 'another disciple', 'known unto the high priest', actually went inside the house with Jesus (John 18^{15}). This disciple must have been fairly well connected to move in high-priestly circles. Moreover, no one would doubt that Levi (or Matthew) the publican would be lacking in mental alertness. It is therefore wrong to assume that, as Jesus came to call sinners and outcasts, this necessarily implies he called the ignorant.

Extremely exacting demands were made by Jesus upon his disciples. The presence of the Kingdom of God means that everything must serve the end of the Kingdom itself. 'He that loveth father or mother more than me is not worthy of me; and he that loveth son or daughter more than me is not worthy of me' (Matt 10^{37}). It is not that no longer should parental or filial love exist, but that they should not hinder the establishment of the Kingdom. Any task or job that one has to do must receive secondary consideration in the light of the demands of the Kingdom. 'And another of the disciples said unto him, "Lord, suffer me first to go and bury my father." But Jesus said unto him, "Follow me: and leave the dead to bury their own dead"' (Matt 8$^{21, 22}$). Here is an echo of the crisis and urgency that has entered into men's lives with the advent of Christ. One can feel some sympathy with the Jewish disciple who is told to 'eat such things as are set before you' (Lk 10^8) when on a missionary journey. The Jewish food laws, their dietary restrictions, were firmly held and respected, but even these had to go for the sake of the gospel.

Acceptance of Jesus would mean rejection and persecution by men. 'They shall lay their hands on you, and shall persecute you, delivering you up to the synagogues and prisons, bringing you before kings and governors for my name's sake' (Lk 21^{12}). Jesus warned them of the reality of the situation. His disciples would receive similar treatment for their adherence to the faith as the Chasidic party (symbolized by lambs in the book of Enoch), violence and suffering. 'Behold, I send you forth as lambs in the midst of wolves' (Lk 10³). Perhaps a mitigating circumstance is the realization that they were only being asked to endure the same rejection that Christ experienced. 'The foxes have holes, and the birds of heaven have nests; but the Son of Man hath not where to lay his head' (Matt 8²⁰). The point of this saying is missed if it is regarded as a quaint saying. Jesus referred to Herod as a 'fox' and the 'birds of heaven' is a common apocalyptic term for the Gentile nations. Therefore Jesus implies that the Roman power and the ruling authorities had made their position secure in the land, but the true Israel was dispossessed. By joining Jesus, one joined the ranks of the dispossessed. Under such conditions could a disciple expect to work.

Yet the total picture is not one of unrelieved gloom. Inasmuch as disciples suffered with Christ, they also participated with him in his glory. Even at that time, theirs was the joy of experiencing membership of God's Kingdom that had been denied to others. 'Blessed are the eyes which see the things that ye see, for I say unto you, that many prophets and kings desired to see the things which ye see and saw them not; and to hear the things which ye hear, and heard them not' (Lk 10²³, ²⁴). They had a security in the knowledge that God's grace was always present to support them in any difficulty: 'Ask and it shall be given you; seek, and ye shall find; knock, and it shall be opened unto you' (Lk 11⁹). In fact they knew that God was aware of their needs before they asked. There were definite rewards for them both in the present and in the future: 'Then answered Peter and said unto him, "Lo, we have left all and followed thee; what then shall we have?" And

Jesus said unto them, "Verily I say unto you, that ye which have followed me in the regeneration when the Son of Man shall sit on the throne of his glory, ye also shall sit upon twelve thrones, judging the Twelve tribes of Israel. And everyone that hath left houses, or brethren, or sisters, or father, or mother, or children, or lands, for my name's sake shall receive a hundredfold, and shall inherit eternal life" ' (Matt 19$^{27-29}$).

Such were the conditions to which disciples in general, and the Twelve in particular, were called. What little we know of the Twelve personally strengthens the view that they were ordinary human beings. As they experience life with Jesus, they can show surprise at what they see. 'And they feared exceedingly and said to one another, "Who then is this, that even the wind and the sea obey him?" ' (Mk 4^{41}). At times they appear remarkably obtuse and fail to grasp Jesus' meaning. 'Whilst crossing the lake, Jesus said, "Take heed, beware of the leaven of the Pharisees, and of the leaven of Herod." And they reasoned one with another, saying, "We have no bread." ' (Mk 8$^{15, 16}$). On several occasions the strong personality of Jesus seems to dwarf the disciples: 'And no man after that durst ask him any question' (Mk 12^{34}). One can feel much sympathy for the group of disciples who stood at the foot of the Mount of Transfiguration and tried to heal the epileptic boy whilst some of the crowd jeered. Later they pathetically inquire of Jesus, 'Why could not we cast it out?' (Matt 17^{19}).

But it was that inner group of disciples, Peter, James and John, selected for special instruction on important occasions, who displayed their human shortcomings most forcibly. The two brothers were nicknamed Boanerges, the sons of Thunder, and their temperament fits the description. When some Samaritans were inhospitable to Jesus, they wanted to provide a quick remedy; 'Lord wilt thou that we bid "fire to come down from heaven and consume them?"' (Lk 9⁵⁴). They appeared very jealous of their office as disciples and one can detect a note of exclusiveness in their observation to Jesus, 'Master, we saw one casting out devils in thy name: and we forbade him, because he followed not us' (Mk 9³⁸). A most blatant example of self-seeking is reported in Mark 10³⁵⁻⁴⁴ when they asked Jesus for a privileged place in the Kingdom of God!

Peter's character is so well known as to need no comment. But this person, chosen by Jesus as the disciples' leader, was the one whom Jesus called 'Satan' and who in the final crisis denied Jesus. None of this has been said to diminish the Twelve's authority or to detract from their importance but to show that they, like us, were subject to human limitations. There is great comfort in the fact that Jesus said to the leader of the disciples, 'Simon, Simon, behold, Satan asked to have you, that he might sift you as wheat: but I made supplication for thee, that thy faith fail not: and do thou, when once thou hast turned again, stablish thy brethren' (Lk 22³¹, ³²).

The work of a disciple was, above all, to proclaim, 'The Kingdom of heaven is at hand' (Matt 10⁷). In its turn this involved bearing witness to the life, death and resurrection of Jesus Christ. As the presence of the Kingdom was incompatible with any of the imperfections of the unregenerate creation, the command to preach was combined with the order to heal the sick and cast out demons (Matt 10⁸). One important aspect of their task was to serve others. Leadership involved service and humility. 'The Kings of the Gentiles have lordship over them, and they that have authority over them are called Benefactors. But ye shall not be so: but he that is the greater among you, let him become as the younger: and he that is chief as he that doth serve. —But I am in the midst of you as he that serveth' (Lk 22²⁵⁻²⁷).

THE DISCIPLES OF JESUS

POINTS FOR DISCUSSION

1. What special instructions did Jesus give his disciples when he sent them out two by two? What dangers would they encounter? What special help would they receive?
2. James and John showed a lack of understanding concerning Jesus' ministry whilst they were travelling through Samaria and when they asked him for a privileged position in the Kingdom. Narrate these events and explain fully what teaching Jesus then proceeded to give them.
3. What evidence is there in the gospels for believing the following:
    (*a*) Some of the disciples were well educated or wealthy
    (*b*) That there were considerably more than twelve disciples
    (*c*) That the disciples often experienced difficulty in seeing what Jesus meant
    (*d*) That the disciples discussed Jesus amongst themselves?
4. Relate the story of Peter's denial of the Lord. How far is it consistent with what we learn of Peter elsewhere in the Gospel? (Durham, 1956.)

12

THE CONFESSION AND THE TRANSFIGURATION

A. The Confession

Throughout the early period of the ministry two activities appear to be occurring alongside one another. Jesus teaches large crowds and in so doing creates a certain amount of tension and opposition amongst 'official' circles. The very size of the crowd following could be a source of embarrassment and hinder Jesus in his work, and he risked being arrested on several occasions. Whilst this popular movement is progressing, Jesus tries to give his disciples some distinct teaching to assist them in their special task. Therefore it is not surprising to find Jesus withdrawing from the demands of the crowd at Caesarea Philippi in order to teach his disciples.

'Who do men say that I am?' (Mk 8:27). The reply shows that Jesus was generally considered to be someone different, like an Old Testament prophet or John the Baptist, a man inspired by God to give a message to the people. As the disciples had been so close to Jesus there can be little doubt that they thought of him as Messiah. Therefore Peter's reply, 'Thou art the Christ' (Mk 8:29), cannot be understood as a sudden revelation or insight into Jesus' person. It is much more likely to have been the first open declaration of the thought, in front of Jesus himself. This is important. What one person says to another about themselves can alter their relationship profoundly. For example,

THE CONFESSION AND THE TRANSFIGURATION 109

the declaration, 'I love you' or 'I would like to enter into a business partnership with you', alters the personal relationships in so far as they become more definite. Peter's confession represents a vow and a personal commitment to Jesus. What, up until this time, had been a possibility now becomes a reality.

That Peter understood the implications of what he was saying is extremely doubtful, as the sequel shows. Jesus proceeded to warn his disciples that his messiahship involved suffering and that anyone following him must 'deny himself, and take up his cross' (Mk 8³⁴). Peter had missed the point when he remonstrated with Jesus that his suffering was not necessary. 'Get thee behind me, Satan: for thou mindest not things of God, but the things of man' (Mk 8³³). This persistent obtuseness of the disciples, the failure to realize the implications of their task, has often struck commentators as odd. What is frequently forgotten is that it is only by *doing* a job that one realizes what it involves. Peter had acknowledged the fact of his relationship with Jesus; what this fact implied in terms of personal denial he did not realize. How could he? This was a novel situation and experience would teach. Jesus himself had often explained that true discipleship meant not only making a verbal assent to Christ's messiahship but also doing something when one saw the demand. Of course the demand is conditioned by accepting Jesus as the Christ. Therefore the disciples had taken the first important step.

The events at Caesarea Philippi have great implications for the future rôle of Peter as the recognized leader of the disciples. 'Thou art Peter; and upon this rock I will build my church; and the gates of Hades shall not prevail against it. I will give unto thee the keys of the Kingdom of heaven; and whatsoever thou shalt bind on earth shall be bound in heaven; and whatsoever thou shalt loose on earth shall be loosed in heaven' (Matt 16¹⁸, ¹⁹). The 'keys' here are similar to those referred to in Luke 11[52], 'the key of knowledge', the ability to let men into the Kingdom of God. Peter is also given the authority to make (bind) or rescind (loose) regulations for the smooth running of the Christian society.

B. *The Transfiguration*

Jesus wanted to leave no doubt in his disciples' minds that he was the Christ. Although Peter had previously told the disciples that Jesus was the Son of God, this lesson was repeated in a different and visually more dramatic manner. The scene was a mountain, a similar mountain to the one upon which God had appeared to Moses and informed him of God's purpose for the Jews. Now Jesus ascends another mountain with Peter, James and John and makes God's will known to them.

At first sight there appears nothing unusual in Jesus' ascent to the mountain, because Mark tells us that he often used to do this in order to pray with the disciples. But each evangelist connects this ascent with the previous confession of Peter by making it occur at a definite period of six, or eight, days after that event. The remarkable change that they saw in Jesus' appearance made it difficult for the evangelists to describe it in everyday language. Luke partly overcomes this difficulty by stating that the disciples were very tired and sleepy but emphasizes that they were fully awake when the transfiguration happened. All of the evangelists stress that this was a spectacular event by distinguishing those elements in the situation which occurred at other crises in Jesus' ministry, namely, his baptism and death. The cloud, and the voice bearing witness to his Sonship are the most obvious elements, and the oblique reference to Elijah remind us that this is a crisis.

The language and the form of this crisis are paralleled in another encounter of God and man, namely, Yahweh meeting with Moses on Mount Sinai. The setting is a mountain; God appears and speaks to man, a God whose shining appearance forces Moses to hide his eyes. Meanwhile there is confusion at the foot of the mountain where Aaron is indulging in idol worship. Jesus finds confusion at the foot of the mountain where his disciples are being ridiculed by the Pharisees because of their inability to cure the epileptic boy. Here then are two revelations of God. Peter realized the connection between the two events when, almost unconsciously, he asked Jesus whether they should build three tabernacles, one for Moses, one for Elijah and one for Jesus. At the mention of the word tabernacle, one is reminded of Moses and the wilderness. Luke strengthens the link between the two incidents by saying that Jesus was talking to Moses and Elijah about his 'exodus', that is, his departure from the world.

Although the language and form of the incident gains inspiration from the past, the significance of the event is set in the future. Jesus is not only the Son of God but also the suffering Christ. He discusses his coming death with Moses and Elijah, and on the way down from the mountain he explains to the disciples the necessity for his suffering. Elijah is a particularly apt choice of an image through which Jesus can explain his suffering. It was Elijah who had a spectacular exit from the world, being taken up in a chariot of fire, never to be seen again. So that the Jews believed that at the critical point of their history, Elijah would return to herald the Messiah. Consequently, when the disciples were beginning to realize that Jesus was the Christ, their minds immediately turned to the tradition of the return of Elijah. The reply, that Elijah had returned, seems a little startling. But it should be borne in mind that John the Baptist was popularly regarded as Elijah.

This was no chance happening; Jesus was fulfilling the scriptures about himself. The Jewish scriptures were divided into three main sections (the Law, the Prophets and the Writings), and the first two sections were the more important. Moses and Elijah were the ideal representatives of the Law and the Prophets, and Jesus was seen discussing his future actions with them. This then was a way of saying that Jesus had divine sanction for the suffering he was about to undergo, and which his disciples had such difficulty in understanding. In the discussion on the way down the mountain, Mark makes much of this point when he reports Jesus explaining 'how it is written of the Son of Man'.

It is important to see the Transfiguration not only as an important event within itself, but also as having a significant place within the framework of the ministry. After this event the predictions of his death become more insistent. Matthew and Mark stress this by repetition, in Mark 9^{31} and $10^{33,\ 34}$ and in Matthew $17^{22,\ 23}$ and $20^{18,\ 19}$. Luke himself makes Jesus more conscious of the destiny he is to fulfil, in Luke 13^{31-35}. A greater awareness of the end has its effect upon Jesus and his disciples. Mark informs us (10^{32}) that Jesus walked ahead alone, in apparent isolation, and that the disciples were amazed and afraid at this change. Some of the more clear-sighted disciples like James and John are worried about their position when Jesus enters his glory, only to be told that they too will share in his suffering, his 'cup' of bitterness. Jesus anticipates the fears of all his disciples and begins to teach them more about their true calling (Lk 9^{57-63}). The emphasis is upon the necessity of service rather than on the privileges of their position. The change in the ministry is subtle but quite definite.

But by far the greatest change after the experience of the Transfiguration was that the messianic secret was out

THE CONFESSION AND THE TRANSFIGURATION 113

in the open. Previously, Jesus, when recognized by individuals or demons, had insisted upon silence, but now he allows himself to be recognized. The first indication of this is on the occasion of his cure of the blind Bartimaeus at Jericho, and the final climax is reached with the crowd's acclamation at the entry to Jerusalem on Palm Sunday. It is not so much that there are predictions like the one at the end of Luke 13, 'Verily I say unto you, "Ye shall not see me until the time come when ye shall say, 'Blessed is he that cometh in the name of the Lord' " ', but that Jesus now presumes that he will be recognized as the Son of God. What a strong assertion of authority there is in Luke 13[34], 'O Jerusalem, Jerusalem, which killest the prophets, and stonest them that are sent unto thee: how often would I have gathered thy children together, as a hen doth gather her brood under her wings, and ye would not.'

POINTS FOR DISCUSSION

1. 'Who do men say that the Son of man is?' 'But who say ye that I am?' What answers were given to these questions? To what teaching of Jesus did these answers give rise? (Ox. and Camb., 1958.)
2. 'The Transfiguration marks a turning point in Jesus' ministry.' Why should this be so? What changes do you detect in the ministry after the Transfiguration?
3. Tell in your own words the story of what happened at Caesarea Philippi. What is the importance of this event in the ministry of Jesus? (S.U.J.B., 1958.)
4. What Old Testament passages are alluded to in the account of the Transfiguration? Comment on the significance of these allusions.
5. 'Peter was the obvious leader of the disciples.' Comment on this with reference to Peter's statements at the Confession and Transfiguration.

The events of the Last Week according to St. Mark's Gospel

SUNDAY	Two disciples obtain a colt (11^{1-3}) Entry to Jerusalem (11^{4-10}) Visit to Temple; the night spent at Bethany (11^{11})
MONDAY	The Fig Tree incident (11^{12-14}) Cleansing the Temple (11^{15-17}) Growing opposition but left city at night (11^{18-19})
TUESDAY	Teaching concerning the Fig Tree (11^{20-25}) Questions: Authority (11^{27-33}); Parable, Wicked Husbandmen (12^{1-12}) Tribute to Caesar (12^{13-17}); Resurrection (12^{18-27}) Greatest Commandment (12^{28-34}) David and Christ (12^{35-37}); Widow's mite (12^{41-44}) Little Apocalypse (13)
WEDNESDAY	Official hostility (14^{1-2}) Anointing at Bethany (14^{3-9}) Judas goes to chief priests (14^{10-11})
THURSDAY	Preparations for the passover (14^{12-16}); Last Supper (14^{17-26}) Gethsemane (14^{32-52}); Interrogation by High Priest (14^{53-65}) Peter's denial (14^{66-72})
FRIDAY	Trial before Pilate (15^{1-15}); Crucifixion (15^{16-39}) Body removed in the evening (15^{42-47})
SATURDAY	Mary Magdalene, Salome, Mary the mother of James buy spices (16^1) — SABBATH
SUNDAY	Stone is rolled away, young man in tomb, Jesus is risen (16^{2-8})

13

THE LAST WEEK

The last week in Jesus' ministry is a crisis. Jesus openly professes himself the Christ and does this at the headquarters of Jewry, so to speak, in Jerusalem. The authorities cannot ignore his claim to be Messiah. Each evangelist stresses that Jesus was Christ by right of fulfilling scripture. But this does not mean that the clash between Jesus and the Jewish authorities was restricted to a narrow and barren discussion upon the meaning of certain Old Testament passages. Jesus, both by his actions as Messiah and by fulfilling the scripture, focuses the attention upon himself. It is no longer a question of accepting or rejecting his teaching but of accepting or rejecting him. The personal animus aroused by the choice he presents to the Jewish authorities becomes critical. How personal this clash was can be judged by the fact that, ultimately, it is Jesus alone who faces the authorities in the Garden of Gethsemane and during the trials.

Sunday

The entry to Jerusalem on Sunday was the act of a King. A monarch entering his capital is publicly acclaimed and it is a time for rejoicing.

'Tell ye the daughter of Zion,
Behold, thy King cometh unto thee.'

But this is a King with a difference because 'he cometh in

the name of the Lord.' This point is especially stressed by all the evangelists in that many messianic passages are quoted to substantiate the claim fully. Isaiah (62^{11}) and Zechariah (9^{11}) had foretold it. He was not coming empty-handed but was prepared to offer them something essential —salvation. This was no earthly warrior riding in pomp but a meek person, riding on an ass. Even the fact that the ass had not been used by man before suggests that it was specially set apart for God's use. (Compare Num 19^2, Deut 21^3, Lk 23^{53}.)

Although the significance of the act is perfectly clear to the evangelists it certainly was not for those who were actually there. Obviously there was some bewilderment when Jesus entered the city and Matthew reports some asking, 'Who is this?' and they were told, 'This is the prophet, Jesus, from Nazareth in Galilee.' For many then, they were merely acclaiming a prophet from the North. Moreover, it does not emerge too clearly who were the people engaged in throwing down their garments and shouting. Mark says 'many' did it; Matthew is a little more specific saying it was 'the most part of the multitude', whereas Luke, in a non-commital way says '*they* spread their garments on the way' but the shouting was done by the 'whole multitude of the *disciples*' because of 'all the mighty works which they had seen'. Whatever one may make of this crowd there is no need to see here a band of Galilean pilgrims entering the city for the feast acclaiming their hero. Admittedly, Jesus has not been reported as having dealings with Jerusalem during his ministry by the synoptic writers but he had such close friends near by (Simon the leper, Mary and Martha, Joseph of Arimathaea) and was so obviously concerned with Jerusalem that it is difficult not to assume that many 'Southerners' were in the crowd too.

The meaning of the statement 'Blessed is he that cometh

from the Lord' is made apparent by the sequel in Mark. Jesus goes to the Temple and views everything until evening. This is not to imply a quiet end to a rather hectic day, nor is it in any sense an anticlimax to the great claims made earlier by the crowds. If Jesus came from God then the first visit he would wish to make in the city would be the place where God was worshipped—namely, the Temple. Mark's dramatic use of inactivity and silence leave many questions to be answered. What impression did the scene at the Temple have upon Jesus? What would he do next? Even when he starts describing the events next day, he keeps us in suspense before giving the answers to these questions.

Luke, however, is much more concerned, at this point, with the reaction of people to Jesus. Some of the Pharisees remonstrated with Jesus to keep his disciples quiet but even if they kept silent the stones would cry out! Luke knew the final outcome of all this enthusiasm—the rejection of Jesus. In so far as Jerusalem rejected Jesus personally, it also rejected the salvation and redemption that he came to offer. In A.D. 70 Jerusalem had taken recourse to another method, to secure its salvation and political autonomy, by revolting against Roman authority. The consequences were disastrous, and Luke, writing after the event, sees the evil days of A.D. 70 as a consequence of Jerusalem's refusal to acknowledge Jesus. All the atrocious horrors, originally contained in a Jewish taunt song (Ps. 137) against her Babylonian conquerors, are turned back upon Jerusalem herself 'because thou knowest not the time of thy visitation'.

Monday

Mark, unlike the other evangelists, allows a day to elapse before Jesus re-enters Jerusalem and cleanses the

Temple. Before describing this event, he inserts an incident commonly called the 'cursing of the Fig Tree'. At first sight, this appears as an irrelevant insertion which raises acute difficulties. Many have seen this action as a miracle which Jesus performs on an innocent fig tree in a rather vindictive and arbitrary manner. Nothing could be further from the truth. Mark uses this event as a key incident to explain later actions of Jesus more fully.

The 'cursing of the Fig Tree' is to be regarded as an enacted parable, similar in many respects to Jeremiah's use of the potter's pot and Ezekiel's rather eccentric activities. The action is intended to convey a message in a direct and spectacular manner. The message here obviously concerns Israel, for she had been compared with a fig tree on many previous occasions (Judges 9^{7-15}, Jer 24^{1-10}). Viewed from a distance Israel gave the appearance of a vigorous and healthy growth, coupled with the prospect of producing worth-while fruit. Yet it was not reasonable to assume that she had already progressed so far in the knowledge of God's will, 'for it was not the season of figs'. Upon closer inspection, the fig tree was discovered to be barren and was merely putting up a show—a façade— and deluding everyone. If the incident is viewed in this way, it provides the connecting link between Jesus' first visit to the Temple on Sunday and the reason for his anger and violent action on Monday.

The meaning of the 'cursing of the Fig Tree' has been obscured because many scholars have emphasized Peter's outburst on the following morning, 'Rabbi, behold, the fig tree, which thou cursedst, is withered away.' Peter, as on many other occasions, has missed the point. It was not that a marvellous miracle had been performed but that what God had said was certain to happen. Jesus then continues, 'Have faith in God.' He wants the disciples to realize that an absolute trust in God's power, to accomplish what he has said he will do, is essential for them. Jesus is demanding not so much a demonstration of his disciples' faith as their knowledge of God's power. All things are possible for God, and the passage in Zechariah 4$^{6-7}$ would happen if God so desired. The whole point of this passage is missed if one assumes that it is possible for a man to move a mountain if he has enough of his own faith in God.

The Fig Tree episode points to two conclusions. Firstly, Judaism is fraudulently pretending to be something that it is not and will be judged. Moreover, God's purposes cannot be thwarted, and He will do what He has stated. Mark is then ready to show how this judgement was executed by Jesus' entry to the Temple and cleansing of it.

The cleansing of the Temple is a fulfilment of the prophecy contained in Malachi 3$^{1\text{ff}}$.

'... the Lord, whom ye seek, shall suddenly come to his temple... And he shall sit as a refiner and purifier of silver, and he shall purify the sons of Levi, and purge them as gold and silver.'

The immediate cause of cleansing is given as the defilement of the Temple area by the money changers and by those selling sacrifices so that the primary purpose of the Temple as a place of worship is hindered. Matthew and Luke, both writing after the destruction of the Temple, do not finish the quotation from Isaiah 56^7 because by that time it was certain that the demolished Temple could no longer fulfil this rôle.

In spite of the brevity of Matthew's description of the incident, he adds two interesting details to the narrative. Jesus healed the blind and the lame within the Temple. This piece of detail seems to be inspired by the story of David in II Samuel 6–8 which states the origin of a popular proverb 'There are the blind and the lame; he cannot come into the house.' Matthew therefore, seems to be saying that now the son of David is in the Temple, none will be excluded, not even the blind and the lame. The other piece of information which Matthew adds is that the crowd shouted, 'Hosanna to the Son of David' which produced an indignant reaction from the authorities. Jesus answers the rebuke from Psalm 8, which in the Hebrew version states,

> 'Out of the mouths of babes and sucklings hast thou established strength, because of thine adversaries, that thou mightest still the enemy and the avenger.'

This recalls the passage in Matthew 11[25] where Jesus thanks God that although his will is hidden from the sophisticated, yet babes understand it. Such a reply demonstrates the certainty that Jesus is fulfilling his Father's will and the possibility of an almost intuitive grasp (by some of the people) of what he was doing.

Tuesday

The authorities at Jerusalem could hardly allow such activities without making a protest, because Jesus' attack upon the established institutions also represented a personal attack upon the upholders of this system. Moreover Jesus' claim to be Messiah was a threat to the position of the normal authorities. Thus, realizing their tenuous position, the Jewish authorities oppose Jesus in argument to try to make the people acknowledge the falsity of Jesus' claims. For this reason many commentators have called Tuesday, 'The Day of Questions'.

'By what authority doest thou these things?' Such was the first radical question. A perfectly legitimate answer would have been that it was the authority that God had given to His Son. But Jesus could not answer in such a forthright manner because his opponents were spiritually blind—they could not see that Jesus was the Son of God. Therefore Jesus deflects the question from himself by asking them, 'The Baptism of John, was it from heaven, or from man?' This question contains the same principle involved in their question, namely, the possibility of a man being divinely inspired. They were embarrassed to give an answer to this point, 'for all verily held John to be a prophet'.

Once the Jewish officials had shown their incapacity to acknowledge God's servants, Jesus presses the point home by telling the parable of the wicked husbandmen, which is intended to show that the Jews have always rejected those whom God has sent to them. Jesus undoubtedly had the passage in Isaiah $5^{1\text{ff.}}$ in mind here, and the whole section is a comment upon the Hebrews' reception of the divine revelation. The details of the story are significant only in so far as they make the main point clearer. At the end of the parable, the audience is required to make a judgement, 'What therefore will the lord of the vineyard do?' When this has been answered, it follows logically that one has a valid reason for the reception of the Gentiles into the early church, for he 'will give the vineyard unto others'. Luke and Matthew go further than Mark by saying that the 'stone' which the Hebrews rejected will become the foundation of this new structure (Ps 118[22, 23]).

The Jewish leaders return to the attack with a question charged with political implications. 'Is it lawful to give tribute unto Caesar, or not?' A simple 'Yes' or 'No' reply was impossible because the one would have alienated the people from Jesus, as all the Jews bitterly resented the levy paid to an occupying power, and the other would have involved Jesus in treasonable practices. Many have seen in Jesus' reply, 'Render unto Caesar the things that are Caesar's, and unto God the things that are God's', an assertion that good citizenship is always perfectly compatible with being a good Christian. Such a principle is extremely dangerous. In recent times one state has demanded that its citizens should persecute and kill the Jews. Our own legal system acknowledges the right of citizens to opt out of certain obligations to the state for the sake of conscience. Jesus is not promulgating a dictum regulating men's actions as Christian members of a state. He is saying that men are quick to admit responsibilities to the state, but they are not willing to concede, with such alacrity, that God also makes certain demands upon them. In fact, the Pharisees and Herodians had their priorities rather confused in asking such a question.

In many ways it was the rulers' worldly wisdom, their ability to look at life from man's point of view and to arrive at reasonable solutions, that divided them from Jesus and made it impossible to accept his point of view. Jesus appeared unreasonable to them because he, having greater knowledge of what God wanted, and, looking at man from God's point of view, arrived at conclusions which did not seem 'fair' or 'reasonable' to them. This divergence of viewpoint is seen in the next encounter with Sadducees over the possibility of a resurrection from the dead and a life after death. The reason they did 'greatly err' was that they applied reasonable human principles to a situation that could only be understood by reference to God's ultimate purpose for man. The life hereafter was not merely an extension of this life, with all its responsibilities and commitments maintained intact. Completely new relationships will be involved. One can sympathize with the Sadducees because, apart from Jesus Christ, there is no single argument for the resurrection of the dead that is reasonable.

If Jesus knew God's will, then he was in a position to give some constructive suggestions to those who were seeking to do it. So far, the questions had prompted negative replies to indicate where the interrogators were at fault. Now the next inquiry, 'What commandment is first of all?', demands a positive reply. This statement is so important that it must be quoted fully.

> 'The first is, "Hear, O Israel: the Lord our God, the Lord is one; and thou shalt love the Lord thy God with all thy heart and with all thy soul, and with all thy mind, and with all thy strength." The second is this, "Thou shalt love thy neighbour as thyself". There is none other commandment greater than these.'

Such an answer is astounding. It is astonishing that Jesus should cite two passages from the Jewish Law, that is, from Deuteronomy 6[4-5] and Leviticus 19[18]. Obviously, then, these are not new decrees made by Jesus but statements from the Jewish law, which he wholeheartedly endorsed. Although these passages are taken from the law they are not legal demands. No one can legally demand love. Law is concerned with rights and duties: love is concerned with personal relationships, in which individual rights have no place and a greater claim than duty is made upon one. Mother or father, husband or wife, cannot legally demand love, for if they did, it would mean an end of the love relationship and it would deteriorate into a legal contract. When your mother asks you to look after your younger brother, put him to bed and to wash up the tea things whilst she goes out to a particular meeting, you do not say, 'What right have you got to ask this?', or 'What am I going to get out of it?' The task is done, in spite of the fact that it may appear unreasonable.

What sort of demand is being made here by Jesus when he says, 'You must love God'? It is of the same order as saying, 'You must eat, or else you will die.' You can call it one of the facts of life. God created you and gave you all that you have. He asks for your love in that you should show a loving attitude towards the rest of his creation, especially other people with whom you come into contact. Jesus himself had said that this was a way of showing your love towards God; 'Verily I say unto you, inasmuch as ye did it unto one of these my brethren, even these least, ye did it unto me' (Matt 25[40]). Jesus himself lived up to this precept by laying down his life for others and thus showed the love of God for man. If we fail to live up to this demand to love God, then we are sinners.

POINTS FOR DISCUSSION

1. Describe Jesus' cleansing of the Temple and the events occurring immediately after this. What significance do these events have in Jesus' ministry?
2. 'Now from the fig tree learn her parable.' Explain fully what you think Jesus meant by this statement.
3. 'Tuesday in the Last Week is commonly known as the Day of Questions'. State three questions Jesus was asked and give his replies to them. Write briefly what he intended his hearers to understand by each of these answers.
4. How did Jesus answer the question, 'Is it lawful to give tribute unto Caesar, or not?' Has his answer any bearing upon the problem of the obedience due by Christians to the State today, especially in lands where the State is non-Christian? (S.U.J.B., 1957.)

14

AN INTERLUDE: THE LITTLE APOCALYPSE

The continuous and connected account of the Last Week seems to be broken in St. Mark's gospel by the intrusion of a discourse which is not at all characteristic of Jesus' teaching. On first reading, this passage (Mk 13) appears to be concerned with a final catastrophe, in which the disciples will participate and suffer. It is a warning of the signs of the end, phrased in the colourful language of Jewish apocalyptic, a revelation of future events which will be accompanied by the strangest phenomena. 'But in those days, after that tribulation, "the sun shall be darkened, and the moon shall not give her light, and the stars shall be falling from heaven, and the powers that are in the heavens shall be shaken"' ($13^{24, 25}$). Yet amongst all this weird imagery is found an element not typical of Jewish apocalyptic, namely, exhortation and advice. The unexpected nature of such an outburst has persuaded many that this is a later fragment of Christian writing, being concerned with a terrible crisis and persecution of the early church, which therefore has nothing to do with the Passion narrative, but has been inserted there for convenience. Such a view does scant justice to Mark's literary abilities and there are many indications from the text itself that, in spite of a remarkable change of style, the Little Apocalypse is an integral part of the events of the Last Week.

This section begins as a private discussion between Jesus and four of his disciples, Peter, James, John and

Andrew, concerning the destruction of the Temple and the catastrophe which is about to overtake the Jews. Affairs would reach this critical stage because of the choice that Jesus had come to offer men. His rejection necessarily involved the destruction of contemporary Judaism. Suffering and persecution would be the portion of those who accepted him, 'And ye shall be hated of all men for my name's sake' (v. 13). Consequently the disasters about to occur and Jesus' rejection are intimately connected. In particular, this section of the gospel has a relevance for what has preceded and what is to come. Jesus had already 'cast the die'. He had entered Jerusalem, confronted the authorities, criticized their leaders and aroused considerable opposition to himself. What was to come was the Passion narrative of the trials, crucifixion and resurrection. Before completing his earthly ministry, Jesus explains to a chosen few the consequences of these actions. Not that this farewell address is intended only for the few, it is through them to the whole church, 'And what I say unto you I say unto all, "Watch"' (v. 37). These are not predictions concerning the future which all have to be fulfilled literally, but prophetic statements, in the Old Testament tradition, which are based upon spiritual insights into the present.

The present facts for Jesus were the events centred around his imminent death. Therefore the evangelist constructed his discourse with those happenings in mind. Jesus warned the disciples of the possibility of being led astray. 'But take ye heed: behold, I have told you all things beforehand' (v. 23). Several times Jesus had prepared them for his suffering and death and had, at the Last Supper, told them that one of their number would betray him. How the events would work themselves out is seen in verse 9, 'But take heed to yourselves: for they shall deliver you up to councils; and in synagogues shall ye be

AN INTERLUDE: THE LITTLE APOCALYPSE 127

beaten; and before governors and kings shall ye stand for my sake, for a testimony unto them.' The parallel with Jesus before Caiaphas and the Council, as well as the scourging after the Roman trial, is obvious. Mark, in a most unaccustomed manner, introduces a fairly rigid time sequence into the last section, with reference to the need for watchfulness. 'Watch therefore: for ye know not when the Lord of the house cometh, whether at even, or at midnight, or at cock-crowing, or in the morning; lest coming suddenly he find you sleeping' (vv. 35, 36). When one remembers the sequence of events to come, Jesus going for the last supper in the *evening*, the betrayal in the Garden of Gethsemane about *midnight*, Peter's denial at *cockcrow*, and the meeting of the council in the *morning*, this chronological choice cannot be accidental on the evangelist's part.

Even if the chronological sequence seems a little strained, there are more direct allusions to the Passion narrative in the discourse. Jesus' statement about the ultimate outcome is reminiscent of his reply to Caiaphas; 'And then shall they see "the Son of man coming in clouds" with great power and glory' (v. 26). A comment upon the last few hours is found in verse 13, 'but he that endureth to the end, the same shall be saved'. What could be more direct than a definite reference to an event which had just occurred? 'Now from the fig tree learn her parable' (v. 28). In the parable of the Fig Tree Jesus had already connected the worthlessness of Judaism, its fraudulent façade, with its destruction. What is about to happen is a continuation of this same process. Perhaps the most difficult statement to interpret in this section, if it is regarded as a later apocalyptic addition, is verse 30. 'Verily I say unto you, "This generation shall not pass away until all these things be accomplished"' However, if the whole passage is interpreted in the light of the Passion narrative itself, then its meaning is abundantly clear.

The ideas outlined above do not preclude the possibility of some historical allusions in the passage. Indeed, Mark was concerned, as was the whole of the early Church, with the imminent return of Christ, the Second Advent. The fact that this has not yet occurred historically is of no moment in this discussion. What is important is that Mark ardently believed the Parousia (the second coming of Christ) to be imminent and interpreted the events of his own time in the light of this belief. For example, a sign of the time is given as, 'But when ye see "the abomination of desolation" standing where he ought not (let him that readeth understand)' the people should flee to the mountains. The 'abomination of desolation' is referred to in Daniel (9^{27}, 11^{31}, etc.), being the heathen idol that Antiochus Epiphanes set up in the Temple at Jerusalem. Whether Mark was making a specific historical reference to Pilate having the legionaries' standards placed in the Temple in A.D. 19, or to Caligula's effort to have his effigy worshipped by the Jews in A.D. 40, is not the most important aspect of the statement. Rather he is implying that in times of trial in the past there have been comparable events accompanying the persecution. The note in parentheses guards against a too literal or historical parallel being taken, in so far as more is meant than is actually stated.

Although Mark 13 does not report the actual words of Jesus, there is every reason for assuming that it expresses his meaning. The whole chapter is a pertinent comment upon Jesus' impending passion which the evangelist interprets against the background of the persecuted church of his day, struggling to bear witness to Jesus Christ.

POINTS FOR DISCUSSION

1. 'But when ye see the abomination of desolation standing where he ought not (let him that readeth understand), then let them that are in Judaea flee unto the mountains.' Explain fully what you understand by 'the abomination of desolation' in this passage.
2. 'The Little Apocalypse (Mark 13) forms an important bridge between the first and second half of the Last Week.' Explain fully from the content of this chapter the reasons for this statement.
3. 'But when these things begin to come to pass, look up, and lift up your heads; because your redemption draweth nigh.' What are 'these things' to which reference is made in this quotation? (Ox. and Camb., 1960.)

15

THE LAST WEEK

Wednesday

The scene is set for the final climax. Judas himself goes to the authorities to arrange to 'deliver up' Jesus. The officials are becoming impatient. Whilst Jesus is having a meal at Simon the leper's house at Bethany, a woman enters and anoints his head with some extremely valuable nard ointment. Some consider this a senseless and wasteful act. 'To what purpose hath this waste of the ointment been made?' (Mk 14⁴). However, Jesus silences the opposition; 'Let her alone; why trouble ye her? She hath wrought a good work on me.' (Mk 14⁶). Surely the evangelist sees in this act another sign, or witness, to the fact that Jesus is the Christ, the anointed one. This unknown woman's intuition had impelled her to make this act of recognition which, in all probability, she did not fully understand. But, within the context of the events, it contains a greater importance in so far as the 'other women' when they bought spices and wished to anoint Jesus were too late. Jesus had risen. Therefore, the pertinent comment by Jesus, 'She hath done what she could: she hath anointed my body *aforehand* for the burying.' (Mk 14⁸.)

Thursday

'With desire, I have desired to eat this Passover with you before I suffer.' Jesus states the importance, the critical nature of this particular Passover. What is going to distinguish this Passover from all others is the fact that it will be so transformed by Jesus that it will include all that has happened and has been of value in past Passovers and yet go beyond them to make the Passover, as such, redundant. One must be cautious and not dismiss the Jewish Passover as unimportant; on the contrary, its value and meaning must be fully appreciated if the course of the events at the Last Supper are to make sense. By analogy it could be said that a caterpillar bears little resemblance to a butterfly, but it would be foolish to ignore the caterpillar, whilst studying the butterfly. Passover became Eucharist; therefore Passover must occupy our attention first.

The Passover was an event in history, when God intervened through the person of Moses to assist the Jews. God took them out of the power of Egypt, saved their first-born from the fate of the Egyptian first-born—death; promised them a fertile land of their own; welded them together as a nation—instead of keeping them as a rabble of slaves. Moreover, the Jews were given the opportunity of knowing God better, because he had made an approach to them through his great deeds and they were told the necessary response that they should make towards God. To use the biblical phrase, God had redeemed them. He had brought them back under his care and established a new relationship with his people.

To mark this different attitude of God towards the Hebrews and the Hebrews towards God, the Jews held an annual feast called the Passover. It included many elements from the first Passover meal. A lamb 'of the first year and without blemish' was sacrificed, unleavened bread was eaten with herbs, the blood was smeared on the door parts, the whole feast was to be conducted with haste, everything prepared as if the participants were about to go on a journey. Here, through the symbolism of a meal, the Jews remember what God did, and is still doing, for them. This special relationship was reinforced in a further symbolic act of Moses, when he ratified the covenant by sprinkling the blood of an ox over the congregation of Israel and the altar (Ex 24⁸). Man and God were related, albeit through the blood of an animal.

As Jesus and his disciples sat at the Passover table, many of these ideas must have been in their minds. Yet this meal had several parallels with the first Passover feast, which the evangelists could only see when they looked back upon the events after the lapse of time. The sacrifice, the lamb 'without blemish' was Jesus; the food, both the bread and the wine, were to gain a new significance when Jesus said, 'This is my body', and 'This cup is the new covenant in my blood.' All the meal had a sense of urgency and haste because 'the hand of him that betrayeth me is with me on the table'.

Before one has progressed very far in discussing the comparison between the Passover and the Last Supper, a difficulty arises. The original feast was an undertaking between God and a whole people, yet, during the Last Supper, it is Jesus as a solitary figure that is prominent. It is true that, initially, the chance of a relationship with God had been offered to all the Hebrew nation, yet because of the nation's stubbornness and inability to make the relationship work, many of the prophets had seen that ultimately only a minority of the nation could enjoy this relationship. The so-called 'doctrine of the remnant' was one possible answer to the fact that many Hebrews did not realize the obligation of their response to God. One prophet had foreseen the possibility of only one person being completely in harmony with God and obedient to his demands. This shattering revelation occurs in Isaiah 53 when he describes the character of Suffering Servant. On this basis, Jesus could be said to represent the nation.

Just as the Passover gains meaning and significance

from the total experience of the Exodus, so the Last Supper must be viewed against the whole complex of the Passion and the Resurrection. Within this context the actual words employed by Jesus in instituting the communion are very important. He took the cup and gave thanks and said,

> 'This is my "blood of the covenant" which is shed for many' (Mk 14[24]).
> 'Drink ye all of it: for this is my "blood of the (new) covenant", which is shed for many unto remission of sins' (Matt 26[28]).
> 'This cup is the new covenant in my blood, even that which is poured out of you' (Lk 22[20]).

Clearly we are intended to see here a reference in the 'shedding' or 'pouring out' of blood to the sacrifice of Jesus upon the cross and that this is a method of ratifying a 'covenant'. Matthew goes a stage further and connects this event with the remission of sins. All of these terms, sacrifice, covenant and sin, have a long history behind them in the Old Testament. Our first concern then must be to try to assess the meaning of these words for a Jew who was a contemporary of Jesus.

Covenant is often defined as a bargain, the agreement or understanding between God and his people, the Jews. Although this is fundamentally correct, it is inadequate. A bargain or agreement implies an understanding between equal partners containing a *quid pro quo*—'you do this for me, and I will do that for you'. The Old Testament Covenant was never this. God was the dominant, controlling force. He chose the Jews; he made certain demands upon them. They could never go to God and demand favours as a right. The concern and care of God for the Jews was based upon a loving relationship commonly translated in the scriptures as 'the loving mercy and kindness' of God. A loving relationship is usually expressed between people who are intimately connected and the closest of these ties is a 'blood kinship'. Therefore, there is a logical link between 'covenant' and 'blood'. It is the intimate blood relationship that produces the care and affection of father for son. Jesus then is here stating two things. Firstly, that God is once more taking the initiative and seeking to establish a new understanding with man through his son. Further, if man is willing to accept this offer, he can attend the meal as a member of a family, God's family. Both of these ideas are expressed in one form of the Holy Communion Service in the prayer, 'We do not presume to come to this thy Table, O merciful Lord, trusting in our own righteousness, but in thy manifold and great mercies.'

Another central thought at the Last Supper was that Jesus was about to be offered as a sacrifice for men. Sacrifice, to modern ears, sounds a little old-fashioned, unnecessary and crude, particularly its association with animals. But to the Hebrew this was not so. Sacrifice implied a joyous occasion when there was a release of life, not a killing. The release of life was the most sacred part of the procedure, in as much as life was the God-given quality, and, by participating in the ritual, one was very close to one's God. Under these conditions, it is easier to appreciate the nature of Christ's sacrifice. The ultimate outcome of that sacrifice was not death, but life, the hope of everlasting life through the resurrection of Jesus Christ. As important as this idea is, it by no means concludes our understanding of sacrifice. The reason for performing the sacrifice is as crucial as the actual act itself. For the Hebrews, the greatest sacrifice of the year was that performed on the Day of Atonement when God forgave men all their sins committed unwittingly, and his presence was symbolized by smearing the blood of the sacrifice upon the lid of the Ark of the Covenant. The actual area on the cover of the Ark for smearing with blood was called 'the mercy seat'. For Matthew, the evangelist of the Jews, the location of God's presence has been transferred from the Ark to the Cross, but the atoning significance remains. It is interesting to note that a modern scholar believes that Paul's meaning in Romans 3[25] is that it was Jesus 'whom God set forth to be a mercy seat'. Another idea connected with the Day of Atonement that Christians seized upon eagerly was that of the 'covering over' of sins through the sacrifice. The association with the cover or lid of the Ark is obvious. This concept of God covering or blotting out sins through Jesus Christ is not as unreasonable as it sounds at first. The root of sin is pride and self-love. Total destruction of this power is involved in commitment to Jesus Christ. Therefore, it is by freely associating ourselves with Jesus and his sacrifice that we pronounce death to our self-centred existence and obtain life through Jesus Christ.

Perhaps a word of warning is needed at this point. Just because the teaching about the nature of Christ's sacrifice is so important in the Church there have been, from time to time, some rather perverted notions about its interpretation. One of these statements relies upon drawing an odd distinction between the wrath of God in the Old Testament and the love of Jesus in the New Testament. The argument runs that God required the death of His son to appease His wrath. Nothing could be less biblical than this. Another bemused way of thinking claims that Jesus on the Cross took upon him all the sins of men, so that, whatever sins they commit now, do not matter, provided they 'believe' in Jesus. The fallacy here is so obvious that no comment is required.

During the course of the Last Supper an important development occurs. Luke reports it as follows: 'And he took bread, and when he had given thanks, he brake it, and gave to them, saying, "This is my body, which is given for you: this do in remembrance of me." ' Once again the affinities between Passover and Eucharist are seen in that both are commemorative services. The injunction is to remember Christ but in doing so one is not just remembering those very important events of the Passion and the Resurrection. Previously the Fourth Gospel concept of Jesus as 'the Word' has been noted so that his life cannot be restricted to the earthly ministry. He was and is and will be. Therefore, in remembering Christ, one is also calling to mind everything that God has done, is doing and will do for men. Eucharist, the thanksgiving, gathers up into itself all that is important for the person committed to Christ.

An astounding feature of this service is that it is performed in fellowship with other Christians around the family table. Those two normal and essential elements of life, of food and drink, the bread and the wine, symbolize the body and blood of Christ. (Amongst Jesus' contemporaries wine would be as common as tea on our table.) The implication is that, just as one needs to eat in order to live physically, one must also feed on Christ to be alive spiritually. Not that there is a gulf between the physical and spiritual aspects of men's activities; both are inextricably interwoven into everyday life. Let it not be thought that this is a ritual enacted for man's individual self-satisfaction. In the Lucan account, this point is made quite clearly, 'And he received a cup, and when he had given thanks, he said, "Take this and divide it amongst yourselves" ' (Lk 22[17]). Men share the cup in a common fellowship.

These events at the Last Supper are of the utmost importance. Passover became Eucharist. By the special command of Jesus Eucharist is performed today. Certain historical events are remembered in the common act of eating and drinking. This meal is not a normal one but is a symbolic act. Not only do those participating remember the sacrifice made by Jesus upon the cross but they are also giving thanks for everything God is doing for them. What began as the remembrance of specific historical events developed into a general thanksgiving for what God is doing at all times for men.

POINTS FOR DISCUSSION

1. Describe the anointing of Jesus by the woman at Bethany. What importance does Jesus attach to this woman's act? How does St. Luke deal differently with the anointing episode in the house of Simon the Pharisee?
2. What happened during the course of the Last Supper in the upper room? Comment briefly upon the meaning of the statements made by Jesus.
3. Give an account of the contention that arose among the disciples at the Last Supper. How did Jesus rebuke them and what promise did he give them? What memorable words did Jesus say to Peter and what was their sequel? (Ox. and Camb., 1960.)
4. Write short notes on the following: Barabbas, Simon of Cyrene, Joseph of Arimathaea, Mary Magdalene and Judas Iscariot.

16

JESUS ON TRIAL

It is a characteristic of judicial inquiries that one assumes that an unbiased and impersonal judgement can be made upon the basis of the evidence and facts presented to a tribunal. Anything that suggests 'graft', collusion, lies, perjury or downright deceit is foreign to the atmosphere of a law court. When Jesus was brought for trial, the evangelists were in difficulty to describe the proceedings because the real fact of the situation for them was that Jesus was the Christ. However strong this conviction may be and whatever 'evidence' may be adduced to support it, neither the conviction nor the evidence is admissible in a court of law. That Jesus was the Christ is not susceptible of legal proof. By the same token it is absurd to attempt to prove legally that X loves Y—primarily because personal relationships involve extra-legal factors.

Confronted with the fact of Jesus' trials before judicial tribunals, the evangelists feel the need to comment on the absurdity of the situation. All emphasize the illegality of the situation. According to Jewish law, at least two witnesses had to agree in evidence before a case could be brought (Deut 19[15]). Initially this is what the authorities tried to do. 'Now the chief priests and the whole council sought witness against Jesus to put him to death: and found it not '(Mk 14[55]). When this failed, 'false witnesses' were called. This most probably involved some form of bribery and certainly contained hearsay evidence, 'We heard him say . . .'. Still no case had been established against Jesus because 'Not even so did their witness agree together'. Caiaphas overcame this legal obstacle by making the whole tribunal itself witness to the fact of Jesus' 'blasphemy' when he answered, 'I am' to the chief priest's leading question, 'Art thou the Christ, the Son of the Blessed?' Further witness was unnecessary: 'Ye have heard the blasphemy': the law was satisfied.

Although the authorities were convinced of the legality of the trial, the evangelists were not. They all question the timing and procedure of the courts. Jesus had been taken by stealth, in the middle of the night, and then taken immediately to the high priest's house (Mk 14^{53}, Matt 26^{57}, Lk 22^{54}). Here a rather unconventional body of elders and scribes held a preliminary hearing. The whole business appears to have been an *ad hoc* affair. So much so that it was not until the following morning that a plenary session of the Sanhedrin met to discuss the findings of the committee. 'And straightway in the morning the chief priests with the elders and the scribes, and *the whole council* held a consultation' (Mk 15^1). Apparently the object of the meeting of the Sanhedrin was to draft official charges against Jesus which would be valid in Roman Law. The Sanhedrin did not itself have the power to pass the death sentence. Yet the effect of a change in the indictment against Jesus, although perfectly understandable from the nature of the courts, leaves the impression of surreptitious intrigue.

A further difficulty in the court room scenes was that Jesus did not defend himself. Apparently there are two major reasons for this. Firstly, the infamy of such trials is obvious to decent folk. Pilate's wife who had small contact with the case reports to her husband, 'Have thou nothing to do with that righteous man: for I have suffered many things this day in a dream because of him.' The spectacle of a murderer being released, instead of Jesus, speaks for itself and for the judicial processes that can produce such a travesty. Secondly, perhaps the evangelists see in Jesus the silent figure of the Isaianic Suffering Servant. 'He was oppressed, and he was afflicted, yet he opened not his mouth: he is brought as a lamb to the slaughter, and as a sheep before her shearers is dumb, so he opened not his mouth' (Is. 53[7]). Jesus breaks the silence, not to defend himself, but only when silence would imply a denial of the Father and the purpose for which He had sent him.

From what the synoptic writers had said earlier, the fact that Jesus was brought to trial need occasion no surprise. Mark states that there was open hostility between Jesus and the Jewish authorities from the beginning of the Galilean ministry at Capernaum, and that at a very early date 'the Pharisees went out, and straightway with the Herodians took counsel against him, how they might destroy him' (3[6]). Evidence is certainly not wanting to demonstrate an open clash with the authorities on the subject of Sabbath observance, the sanctity of the Temple, the contemporary Pharisaic attitude, let alone such barbed parables as the Wicked Husbandmen. Any casual observer in the Temple area, overhearing Jesus' comments to his disciples, would have been offended by several of his remarks. Therefore it is pointless to attempt to make a moral issue of the trials and to dismiss the Jews as wicked and lawless men. They were not. Jesus was demanding from them, by his very presence, a radical change of attitude towards most of the things they considered important. They were unable to do this because they did not trust him: they were unwilling to associate themselves with him. When put to the test, the leader of the disciples, Peter, failed in this particular respect, when he denied Jesus! The demands Jesus made, and is making, upon man are radical. It was a source of bewilderment to the early Church that all decent men did not embrace Christianity.

JESUS ON TRIAL

Many of those Jews who tried Jesus were decent and respectable and so perhaps the evangelists are implying in their account of the trial that more than decency and respectability are demanded of a Christian.

The trial before Pilate was a political affair. Jesus was arraigned upon an indictment of being politically undesirable and indulging in treasonable and seditious practices. 'We found this man perverting our nation, and forbidding to give tribute to Caesar, and saying that he himself is Christ a King' (Lk 23^2). All of this has an element of truth in it, in so far as Jesus had claimed an authority over and above that of any man. In particular he had violently criticized the central Jewish institution, the Temple, and had actually admitted in cross examination to being the Christ. However, none of Jesus' claims had specific political aims. Jesus was not interested in altering any of the existing political régimes as such, but of altering the way in which men behave within those systems. Pilate was aware of this situation and at a loss to understand the violence of the Jewish demand for the death sentence, 'Why, what evil hath this man done? I have found no cause of death in him.' Some light is thrown upon the course of events by Matthew saying 'For he knew that for envy they had delivered him up' (Matt 27^{18}). In an attempt to break the deadlock, Pilate sent Jesus to Herod.

The Herodian trial was not a complete fiasco. Herod was 'exceeding glad' to see Jesus and took considerable care over his examination: 'and he questioned him in many words'. From what we know of Herod's career, he was no fool. What he wanted was actual proof of Jesus' claims, concrete evidence. Luke emphasizes his desire for proof; 'he hoped to see some miracle done by him' (23^8). Nevertheless, in spite of the taunts and the mocking, Herod sent him back to Pilate, with a statement that he was guiltless of the charges preferred and should not suffer the death penalty (Lk 23[15]).

The Jews insisted and the crucifixion was inevitable.

POINTS FOR DISCUSSION

1. Write a few sentences to explain the most important points contained in the following statements:
    (*a*) 'This man said, "I am able to destroy the temple of God and to build it in three days".'
    (*b*) 'What further need have we of witnesses? Ye have heard the blasphemy: what think ye?'
    (*c*) 'Have thou nothing to do with that righteous man; for I have suffered many things this day in a dream because of him'.
    (*d*) 'And when he knew that he was of Herod's jurisdiction, he sent him unto Herod.'
2. Was there any truth in the charges made against Jesus before Pilate? Give as much evidence as possible from the Last Week.
3. What part did Pilate play in the trials of Jesus? Give a picture of Pilate's character as it emerges in these events.
4. How do the evangelists attempt to show that the arrest and trials of Jesus were illegal?
5. The Pharisees, the Sadducees, the common people, and Pilate were each in part responsible for Jesus' death. Explain why each wished him to die. (S.U.J.B., Summer 1959, adapted.)

17

THE RESURRECTION OF JESUS

Just as Jesus Christ entered our world of time and space in a miraculous manner, being born of a virgin, he made his exit in a miraculous manner, by being resurrected and ascending into heaven. Once more the evangelists were faced with the dilemma of the Son of Man and the Son of God. Jesus of Nazareth, the most human of men, was also uniquely God's anointed, the Christ. As man, one would expect him to die; yet, as Son of God, he had no beginning and has no end. Here is the crucial miracle. It presents philosophical problems of the greatest complexity and scientific problems of the greatest magnitude. Yet, when all is said and done, it is a unique event occurring in history. The problem for the evangelists was to present the unique as intelligibly as the normal.

Their method was to present the facts. On two of these they all agreed. The tomb was empty. Jesus appeared in bodily form to his disciples. A first reaction to these statements is one of incredulity and doubt. It is to the evangelists' credit that they report these normal human reactions faithfully. Once the fact had been revealed, the women fled from the tomb, 'for trembling and astonishment had come upon them: and they said nothing to any one: for they were afraid' (Mk 16[8]). When they recovered their composure and finally did report the matter to the disciples, Luke openly states they were laughed at. 'And these words appeared in their sight as idle talk: and they disbelieved them' (Lk 24[11]). Another typical reaction was

Peter's impulse to go and see for himself and still he did not appear convinced. 'But Peter arose, and ran unto the tomb; and stooping and looking in, he seeth the linen cloths by themselves; and he departed to his home, wondering at that which was come to pass' (Lk 24[12]).

Jesus' appearances cannot be described in terms of hallucination or group hysteria. The only aspects of the appearance of Jesus that are considered strange in the synoptic gospels are the suddenness or unexpectedness of the visitation, and, on the road to Emmaus, the fact that the two disciples did not recognize Jesus; 'But their eyes were holden that they should not know him' (Lk 24[16]). There is nothing untoward in this in so far as a similar inability to recognize Jesus in moments of fear had occurred during his earthly ministry (*cp*. Mk 6[49] and Matt 24[26]). What the evangelists do stress is the normal body of the resurrected Jesus. It could be seen, touched, examined, and Jesus ate with his disciples. ' "See my hands and my feet, that it is I myself: handle me, and see; for a spirit hath not flesh and bones, as ye behold me having." And when he had said this, he showed them his hands and his feet. And while they still disbelieved for joy, and wondered, he said unto them, "Have ye here anything to eat?" And they gave him a piece of a broiled fish. And he took it, and did eat before them' (Lk 24[39]).

Another possibility was that Jesus was not resurrected but that the disciples had removed the body themselves! This eventuality is not overlooked by Matthew, who reports that many people believed this story even at the time he was writing the gospel. The reason for such a false rumour being spread was the fact that the Roman guard had witnessed the opening of the tomb and the fact that it was empty. Fearing their superiors, they had reported the matter first to the Jewish authorities, who then bribed them to put about the story of the disciples' theft of the body. The Jews promised they would straighten everything out with the military authorities, if charges of negligence were preferred against the soldiers (*vide* Matt 28[11ff.]). Even if one does not believe Matthew's evidence, it would involve assuming that the disciples were unscrupulous men who not only deceived others but were foolish enough to deceive themselves and live a life, for many years, dominated and controlled by the very fact of the resurrection. This is manifestly absurd.

Here again, as in so many sections of the gospel, violence is done to an individual occurrence if it is treated in isolation. Luke, in two accounts of resurrection appearances, stresses the importance of the event as fulfilment of scripture. ' "Behoved it not the Christ to suffer these things, and to enter into his glory?" And beginning from Moses and from all the prophets, he interpreted to them in all the scriptures the things concerning himself' (Lk 24[26, 27]). Therefore, the Resurrection must be set within the total context of the scriptures. This does not mean just those Old Testament passages which refer to the raising up of God's anointed, but the significance of life, and a life-after-death seen in terms of God's relationship with man. One very important aspect of this problem is stated in Genesis in the Adam story. God had created man in his own image, as the highest, controlling part of His creation. A natural harmony was to exist within this creation, between man and God, and man and the animals. Something went wrong. Man disobeyed God. Because of this disobedience, death was the punishment. But the Christ came and gave man another chance. He was obedient to God's will; he showed man the way to an intimate communion with God. There was the possibility of a new creation through Jesus Christ; a new harmony became a reality; death was no longer the limiting factor in man's existence. Life-after-death was a logical corollary. The story of Adam is not literally or historically true but it expresses truths about human nature that are borne out by experience.

Another aspect of the same problem is worked out in the book of Job. God is a righteous God, for so the prophets had taught. He demands decent, upright conduct. However, here on earth the good and bad exist side by side and the bad appear to get the better of the bargain. The wicked prosper, the righteous suffer. If God is true to his own righteous character, the good must obtain some reward. The answer the author of Job found was in the future in a life-after-death.

'But I know that my redeemer liveth, and that he shall stand up at the last upon the earth: and after my skin hath been thus destroyed, yet from my flesh shall I see God: whom I shall see for myself, and mine eyes shall behold, and not another' (Job $19^{25\mathrm{ff.}}$). Jesus Christ provided an answer to this moral problem. It is in this sense, and in so many other cases, that Jesus fulfilled the scripture by his resurrection.

The resurrection of Jesus is not a matter that can be dismissed lightly. One cannot ignore its implications. 'I do not believe it.' That is no answer, not even the beginning of one. Disbelief, like belief, needs some rational grounds or proof. A denial of the Resurrection requires a jettisoning of so much that is valid and real in life itself. One must inevitably feel sorrow and sympathy for those who deny the possibility of such a fact. The promise of resurrection is not merely a pious hope that makes the thought of death bearable; its effects are felt here and now in life. This earthly life gains a new perspective, fear is removed and fresh insight can be gained about the purpose of the daily round. Jesus of Nazareth, the Christ, is very significant today, for, as he said, 'I came that they may have life, and may have it more abundantly' (John 10^{10}).

POINTS FOR DISCUSSION

1. What evidence do the evangelists give for the resurrection of Jesus? Comment on any substantial differences you have discovered in the various accounts.
2. Write an account of the events which took place on the morning of the first Easter Day as recorded by St. Matthew. (Ox. and Camb., 1958.)
3. 'But we hoped that it was he which should redeem Israel.' When was this statement made? What does it mean? Were they given a satisfactory explanation so as to allay their fears?
4. What arguments could you put forward to doubt the resurrection of Jesus? What differences would these arguments make to your attitude towards your own life?
5. What does St. Mark tell us about the Resurrection of Jesus Christ? Why does the Revised Version leave a blank space before the accounts of his appearances after the Resurrection? (S.U.J.B., 1957.)

INDEX

Adam, 20, 37, 146–7
Anointing at Bethany, 130

Baptism, 31, (Jesus') 34–6
Barnabas, 16
Bartimaeus, 56, 113
Beatitudes, 94
Beelzebub, 42

Caiaphas, 103, 127, 139
Coin in fish's mouth, 61
'Corban', 51
Costly pearl, 82, 102

David, 24–5, 120
Day of Atonement, 134–5
Day of Lord, 75ff
Dives and Lazarus, 65, 71–2

Elijah, 32, 37, 58, 110ff
Emmaus, 101, 145
Essenes, 31

Feeding of five thousand, 58–9
Fig tree, 61, 118ff, 127

Galilee, 40ff
Garden of Gethsemane, 27, 38, 127
Good Samaritan, 63, 64, 67–9
Great feast, 65, 79

Herod, 31, 32, 104, 105, 141
Hid Treasure, 82

Jairus' daughter, 9, 56–7
James and John, 27, 35, 81, 102, 105–6
Jerusalem (A.D. 70), 11, 16, 25, 30, 40, 113, 115–47
John the Baptist, 30–2, (disciples of) 55, 102, 108, 111, 121

King going to war, 82

Labourers in the Vineyard, 63, 82
Last Supper, 10, 126, 130ff

Mary Magdalene, 21, 42
Mary Mark, 16
Mary (mother of Jesus), 20
Messiah, 24, 26, 27, 30, 31, 32, 65, 91, 94, 108ff, 115
Moses, 37, 58, 72, 96, 102, 110ff, 131ff, 146
Mustard seed, 78
Mystery of Kingdom of God, 66–7, 76

Nazareth, 20, 40

Old Wineskins, 82

Passover, 130ff
Patched garments, 82
Paul, 14, 28, 69, 88, 135
Peter, 16, 27, 38, 61, 103, 104, 106, 108ff, 118, 145
Pharisees, 31, 47–53, 94, 98, 105, 111, 122, 140
Pilate, 128, 141
Prayer, 90–1
Prodigal Son, 69–70, 82

Q (Quelle), 14–15

Sabbath, 43–4
Sadducees, 47, 52–3, 122
'Salt' sayings, 83–4
Sanhedrin, 52, 95, 139
Satan, 37, 38, 109
Sermon on Mount, 59, 86–98
Seventy, 11, 77, 102
Sheep and Goats, 65
Sick of the Palsy, 54–5
Simon the Pharisee, 21, 41
Son of Man, 20ff, 34, 105
Sower, 44, 65, 78
Stilling of Storm, 45, 60
Strong man despoiled, 83
Suffering servant (Isaiah 53), 10, 17, 35, 132, 140
Syro-phoenician woman, 11

Tares, 78
Temple, 21, 38, 44, 54, (veil) 58, 69, (A.D. 70) 81, 95, 117ff
Temptations, 36–8
Ten virgins, 79
Theophilus, 17
Tower builder, 82
'Traditions', 49–50
Transfiguration, 13, 105, 110ff
Twelve 11, 91, 101ff,

Unjust steward, 71
Unmerciful servant, 91

Walking on the water, 60
'We' passages, 17
Wicked husbandmen, 63–4, 14
Widow of Nain, 9
Woman with issue of blood, 57

Zacchaeus, 21, 41